Drainage of development sites – a guide

R Kellagher

Published jointly by

CIRIA sharing knowledge ■ building best practice

HR Wallingford

CIRIA
Classic House
174–180 Old Street
London EC1V 9BP
Tel: +44 (0)20 7549 3300
Fax: +44 (0)20 7253 0523
Email: enquiries@ciria.org
Web: www.ciria.org

Address and Registered Office: HR Wallingford Ltd
Howbery Park
Wallingford
Oxon OX10 8BA
Tel: +44 (0)1491 835381
Fax: +44 (0)1491 832233
Registered in England No. 2562099. HR Wallingford is a wholly owned subsidiary of HR Wallingford Group Ltd.

SUMMARY

The guide is intended to assist all those involved with the foul and surface water drainage of development sites. It is specifically aimed at developments in the UK based on national requirements and international best technical practice. It provides guidance:

- on the approach needed to obtain Town and Country Planning Act (T&CPA) consent
- to developers and consultants on current good engineering practice for design of drainage and sewerage for new sites
- on the issues affecting site drainage
- on hydraulic-related engineering issues
- to key industry documents.

An important output of recent HR Wallingford research is a proposed new methodology for calculating site storage.

It is intended that general engineering practitioners, developers and architects will use this document as a first point of reference for guidance and information on all aspects related to the hydraulics of site drainage. Sites range from small suburban developments to large industrial estates, each having specific features that require particular attention.

Drainage of development sites – a guide

Kellagher, R

 ISBN 0-86017-900-1

HR Wallingford Report SR 574, job number MBS 0403. CIRIA publication X108.

DETR reference CI 39/5/108. This document also constitutes Environment Agency R&D technical report W264.

Keywords
flooding, land use planning, sewerage, site development, SUDS

Reader interest	**Classification**	
Drainage system designers, house-builders, developers, planners, architects, local authorities, regulatory bodies – building, drainage, planning.	AVAILABILITY	Unrestricted
	CONTENT	Technical guidance
	STATUS	Committee-guided
	USER	Developers, architects, engineers, regulators

ACKNOWLEDGEMENTS

This guide has been brought to the industry by HR Wallingford with financial support from the DETR and the Environment Agency and expert guidance from a steering group drawn from all parts of the water industry. It has been clear for some time that there are several aspects of site drainage that give rise to confusion and uncertainty. They include roof drainage, site pipe design, both foul and surface water, and site storage, as well as sustainable drainage methods. All are covered by this guide, which aims to help the developer or general practice engineer take the correct approach when designing site drainage. The guide is specifically aimed at developments in the UK based on national requirements and international best technical practice.

Author

Richard Kellagher is a chartered civil engineer specialising in urban drainage computational modelling and research into sustainable drainage.

Experience in supporting both large and small developers with their drainage issues led to an awareness for the need for a site drainage guide together with research into related matters, such as stormwater drainage, which became a major issue in the 1990s. He has been responsible for several research projects on sustainable urban drainage systems and has assisted the Environment Agency in providing easy-to-use guidance on stormwater requirements for new developments.

This project was led by Richard Kellagher of HR Wallingford with contributions from Richard May and Manuela Escarameia. Assistance and guidance was also provided by the following industry experts:

- Environment Agency — B Winter
- Ebara UK Ltd — S Russell
- Thames Water Utilities Ltd — D Ridgers
- AMEC Capital Projects — G Snarr
- Whitby Bird & Partners — N Price
- Institute of Clerks of Works — D McMullan.

CONTENTS

Tables

Figures

GLOSSARY

Adoption of sewers	The transfer of responsibility for the maintenance of a system of sewers to a sewerage undertaker.
Aesthetic pollution	Solid sewage-related materials that are visible but create little environmental impact.
Antecedent conditions	The condition of a catchment before a rainfall event.
Antecedent precipitation	The relevant rainfall that takes place before the point in time of interest.
Antecedent precipitation index	Expressed as an index determined by summation of weighted daily rainfalls for a period preceding the start of a specific event.
Anti-flooding device	A device specifically designed to be installed in gravity drains or sewers to prevent backflow from a sewer towards a property or group of properties.
Areal reduction factor	A factor applied to point rainfall depths or intensities to generate values applicable to an area.
Attenuation	The reduction in peak discharge of a flood wave accompanied by an increase in duration of increased flow.
Balancing pond	A pond constructed for the purpose of temporary storage of stream flow or surface runoff, which releases the stored water at a controlled rate.
Base flow	Sustained or dry-weather flows not directly generated by rainfall. It commonly constitutes flows generated by domestic and industrial discharge and also infiltration.
Best management practices	Structural and non-structural measures used to store or treat urban surface water runoff to reduce flooding, remove pollution and provide other amenities.
Brownfield site	Redevelopment of a site, often associated with pollution issues.
Catchment	A defined area, often determined by topographic features or land use, within which rain will contribute to runoff to a particular point under consideration.
Cavitation	The process of implosion of air in water that is a function of high velocities, which causes damage.
Colebrook-White equation	An empirical equation relating flow to roughness and gradient of the conduit and the viscosity of the fluid.
Collection system	In wastewater, a system of conduits, generally underground pipes, which receives and conveys sanitary wastewater (domestic and/or industrial) and/or stormwater.
Combined network	A sewer network that collects rainfall from impervious surfaces and foul water from domestic and industrial sources.
Combined sewage	Sanitary or foul sewage mixed with surface water. Also referred to as storm sewage.
Combined sewer	A sewer intended to receive both surface runoff and wastewater (domestic and industrial).
Combined sewer overflow (CSO)	A structure on a combined or partially separate sewer system that allows flows above the sewer pass-forward capacity to be discharged to another sewer, a stormwater retention tank, a watercourse, or to another disposal point.
Consented discharges	Term used in the UK for discharges meeting the conditions imposed by the appropriate public authority for potentially polluting flow to a watercourse or into the ground.
Contributing area	The area of the catchment that contributes storm runoff directly to the sewerage system.
Control structure	A hydraulic device to limit the rate of the flow.

Culvert	A covered channel or pipeline (defined by the Highway Agency as wider than 900 mm).
Depression storage	Natural depressions on the surface of the ground that need to be filled by rainfall before runoff can take place.
Design storm	A synthetic rainfall event of a given duration and return period. It has been derived by statistically analysing a historical series of rainfall events for a specific location.
Detention basin	A basin constructed for the purpose of temporary storage of stream flow or surface runoff, which releases the stored water at controlled rates.
Detention tanks (balancing tanks)	Tanks constructed in a sewerage system to store a volume of water temporarily during peak flows (see off-line and on-line tanks).
Discharge	The volume of liquid flowing through a cross-section of conduit per unit of time.
Discharge coefficient	A coefficient, derived by experiment, applied in a formula by which the theoretical discharge of a fluid through an orifice, weir or nozzle can be correctly calculated.
Domestic (foul) wastewater	Wastewater from household services including outflows from sinks, toilets, washing machines etc.
Drain	A pipeline, usually underground, designed to carry wastewater and/or surface water from a source to a sewer; a pipeline carrying land drainage flows or surface water from a highway.
Drainage	A collection of pipes, channels and other engineering works designed to convey stormwater away from a built-up environment.
Dry-weather flow (DWF)	The flow of wastewater in a sewer during dry weather. The flow consists of sewage and infiltration.
Effluent	Wastewater or other liquid, partially or completely treated, or in its natural state, flowing out of a pipe or treatment plant.
Erosion	Detachment and movement of soil or sedimentary deposits by the flow of water, such as over the ground surface or in a pipe or channel.
Eutrophication	The progressive enrichment of surface waters, particularly non-flowing bodies of water, such as lakes and ponds, with dissolved nutrients, such as phosphorus and nitrogen compounds, which accelerates the growth of algae and higher forms of plant life.
Evaporation	The drying out process of the ground surface, which constitutes a minor part of the losses taken account in rainfall-runoff loss models.
Event (rainfall)	Single occurrence of a rainfall period before and after which there is a sufficient dry period to define its effect on the sewer system.
Extreme event	Single occurrence of an event that is likely to occur very infrequently (eg long drought or big storm etc).
First foul flush	The initial discharge of active sediments and pollutants generally higher than the average concentration of pollutants caused by rainfall.
Flap gate	A gate that opens to let water out but prevents water entering back into a system.
Flood storage pond	A pond constructed for the purpose of temporary storage of stream flow or surface runoff, which releases the stored water at controlled rates.
Flood studies report	Landmark report in UK for catchment hydrology (Institute of Hydrology, 1975).
Flow regime	Typical variation of discharge of a waterway usually over an annual or seasonal period.
Foul sewage	Waterborne waste of domestic or industrial origin excluding rainwater and surface water.

Foul system	A drain or sewer system that has been designed to carry only foul sewage.
French drain/filter drain	The use of a granular trench filled with stone to convey and infiltrate stormwater runoff.
Frequency	The number of occurrences of a certain phenomenon per unit time.
Gradient	The angle of inclination (of pipe), which dictates its capacity and velocity of flow.
Gravity system	A drain or sewer system where flow is caused by the force of gravity and where the pipeline is designed normally to operate partially full.
Greenfield/greenfield site	New development, usually at the periphery of existing urban areas. This creates increased rainfall-runoff and has an impact on existing sewer systems and watercourses.
Greywater	Wastewater from baths and hand basins; sometimes also taken to include wastewater from washing machines and kitchen sinks.
Gross solids	Solids, usually organic in nature, either floating, suspended or deposited, which have a polluting effect on the receiving water. Often restricted to visible solids with one dimension greater than 25 mm.
Groundwater	Subsurface water occupying the saturation zone from which wells and springs are fed. In a strict sense, the term applies only to water below the water table.
Gully	A structure to permit the entry of surface runoff into the sewer system. It is usually fitted with a grating and a grit trap.
Head-discharge	The relationship between a discharge rate and the water level causing that discharge.
Highway	Any road, track, bridleway or public footpath in private or public ownership that is not associated with an individual property.
Highway drainage system	A drain or sewer system constructed for the purpose of draining a highway.
Hydraulic analysis	Assessment of the hydraulic behaviour of a system. Simulation hydraulic modelling of a sewerage network to determine its performance.
Hydraulic capacity	The maximum flow that a pipe of given dimensions, slope and roughness can carry (often quoted as pipe-full capacity, which is a little less than the maximum capacity).
Hydraulic performance	The measure of the capacity of the system or part thereof.
Hydraulic simulation	The computational process carried out by a computer model to analyse the behaviour of a system (sewer network) due to an external influence (rainfall).
Hydrograph	A graph showing, for a given point on a stream or conduit, the discharge, stage, velocity, available power or other property of water with respect to time.
Impermeable surface	Surface that resists the infiltration of water. Usually a measure of roof and road surfaces in simulation modelling.
Industrial discharge	Outflow from an industrial unit, which varies enormously depending on the processes carried out in the factory.
Infiltration	(a) The unintended ingress of groundwater into a drainage system (also termed parasitory flow in some countries) (b) the introduction of rainwater runoff into the ground.
Infiltration (to sewer)	The ingress of groundwater into a drain or sewer system through defects in pipes, joints or manholes.
Inflow	Flow which enters the sewer; this can be generated by rainfall or an industrial discharge or other particular connection.

Initial loss	In hydrology, rainfall preceding the beginning of surface runoff. It includes interception, surface wetting and infiltration.
Inlet	(1) A connection between the catchment area and a drain or sewer for the admission of surface or stormwater. (2) A structure at the entrance end of a conduit. (3) The upstream end of any structure through which water may flow.
Inspection chamber	A structure that offers access to the drain or sewer for servicing by means of equipment remotely operated from ground level; no personal access.
Intensity-duration-frequency	The relationship between rainfall intensity (amount per unit of time), rainfall duration (total time over which rainfall occurs) and frequency (return interval) at which the specific intensity-duration relationship is expected to recur.
Interception	The process by which rainfall may be prevented from reaching the ground, for example by vegetation.
Internal drainage boards	These manage ordinary watercourses in areas known as internal drainage districts.
Invert	The floor, bottom or lowest portion of the internal cross-section of a closed conduit.
Land use	Catchments zoned based on economic, geographic or demographic use of land, such as residential, industrial, agricultural, commercial.
Lateral	A private drain or sewer that carries drainage flows from a property to a public sewer.
Lloyd-Davies method	An adaptation by Lloyd-Davies of the Rational Method for storm drainage design.
Major system	In the context of major and minor drainage, this refers to the route followed by storm runoff when the minor system is either inoperative or inadequate. It generally refers to roads and major above ground drainage channels.
Manhole	A structure that provides access for personnel to the drain or sewer for servicing.
Manning's equation	An equation developed by Manning to relate flows in conduits to their size, shape, the gradient and the conduit roughness.
Minor system	The drainage pipes, roadway channels, enclosed conduits and roof connections designed to convey runoff from "normal" storms, to eliminate or minimise inconvenience in the area to be developed. See **Major system**.
Misconnection	An incorrect connection of an inlet or drain to a drain or sewer that is not designed to carry that element of flow (eg foul sewage entering a surface water system or surface water entering a separate foul system).
Model	A series of mathematical equations in a computer developed and used with the aim of replicating the behaviour of a system.
Modified Rational Method	A modification of the Lloyd-Davies method introduced by the Wallingford Procedure whereby the coefficient of runoff was split into two entities (HR Wallingford and Institute of Hydrology, 1981b).
Monitoring	The procedure of measuring effluent characteristics such as flows or pollutants by means of instruments.
Muskingum-Cunge routeing method	A method of routeing flows in channels and pipes, first applied on the Muskingum River in the USA and subsequently modified by Cunge.
Network	In the context of sewers, a collection of connected nodes and links, manholes and pipes.
Off-line tank	Detention tank that is off the normal path of flow in a network, which comes into operation during periods of high flows.

On-line tank	A detention tank through which the flow of sewage is normally conveyed.
Orifice	A constriction in a pipeline to control the rate of flow.
Outfall	The point, location or structure where wastewater or drainage discharges from a pipe, channel, sewer, drain or other conduit.
Overflow	The intentional or unintentional discharge of sewage to the environment before it has been treated.
Overflow weir	Any device or structure over which any excess water or wastewater beyond the capacity of the conduit or container is allowed to flow.
Overland flow	The flow of water over the ground or paved surface before it enters some defined channel or inlet, often assumed to be shallow and uniformly distributed across the width.
Peak discharge	The maximum flow rate at a point in time at a specific location resulting from a given storm condition.
Peakedness	A measure of the sharpness of a rainfall profile; that is, the ratio of the maximum to the mean rainfall intensity.
Peaking factor	The multiple of dry-weather flow used for design of pipe sizes and gradients.
Percentage runoff	The percentage of the rainfall volume falling on a specified area that enters the stormwater drainage system.
Percentile	The percentage of occurrences within a stated range; also applied to rainfall profiles (see **Peakedness**).
Pervious surface	A type of ground surface that allows infiltration of water, although some surface runoff may still occur.
Point rainfall	Rainfall rate at a location, in contrast to the average for the region or surrounding area.
Pollutant	Dissolved or particulate material washed into and through sewers. When discharged into receiving waters, pollutants cause an adverse environmental impact.
Pollution	The addition to a natural body of water of any material that diminishes the optimal use of the water body by the population which it serves, and that has an adverse effect on the surrounding environment.
PR equation	Usually refers to the Wallingford Procedure runoff equation (HR Wallingford and Institute of Hydrology, 1981a).
Primary treatment	The first major treatment in a wastewater treatment facility, usually sedimentation.
Private sewer	A sewer for which responsibility is not vested in the sewerage undertaker. Generally it is collectively owned and maintained by the owner(s) of the building(s) it serves.
Public sewer	A sewer for which responsibility is vested with the sewerage undertaker to maintain it.
Pumping station	A structure containing pumps and appurtenant piping, valves and other mechanical and electrical equipment for pumping water, wastewater or other liquids.
Rainfall intensity	Amount of rainfall occurring in a unit of time, generally expressed in millimetres per hour (mm/h).
Rainfall profile	A series of values of rainfall intensity varying with time; a rainfall event is referred to as a hyetograph.
Raingauge	An instrument used to measure and record the amount of rainfall at an allocated location.

Rational Method	A simple method, used throughout the world, for calculating the peak discharge in a drainage system for pipe sizing.
Reach (river)	A stretch of river between two points, often used where the river characteristics are similar.
Receiving waters	Water body (river or lake) that receives flow from point or non-point sources such as combined sewer overflows.
Regulator	(1) A structure installed in a sewer, conduit or channel to control the flow of water or wastewater at an intake, or overflow or to control the water level along a canal, channel or treatment unit. (2) The term used in UK to refer to the Environment Agency and OFWAT due to their legal involvement in controlling water companies.
Reservoir storage	The phenomenon by which a volume of flow is stored temporarily on a surface or in a length of pipe or channel as the depth and rate of flow increase; the storage is depleted after the peak of the storm passes.
Retention pond	A pond constructed for the temporary storage of surface water runoff, which releases the stored water at controlled rates.
Return period	The reciprocal of the average annual probability of exceedence of a specific flow value or event.
Runoff	Water from precipitation that flows off a surface to reach a drain, sewer or receiving water.
Runoff coefficient	The proportion of total rainfall that appears as total runoff volume after subtracting depression storage, infiltration and interception.
Saint Venant equation	An equation developed in the 19th century by a French mathematician, which takes account of all the physical processes of fluid flow such as momentum and inertia to calculate depth for gradually varying flow states.
Screen	A device with openings, generally of uniform size, used to retain or remove suspended or floating solids in flowing water or wastewater.
Scumboard	A board or plate that dips below the top water level to retain scum and other floatables.
Sediment concentration	The ratio of the weight of the sediment in a water-sediment mixture to the total weight of the mixture. Sometimes expressed as the ratio of the volume of sediment to the volume of mixture.
Sediment transport	The movement of solids transported in any way by a flowing liquid.
Sedimentation	The process of deposition and consolidation of suspended material carried by water, wastewater or other liquids, by gravity.
Self-cleansing (velocity)	The minimum velocity in sewers necessary to keep solids in suspension, so preventing their deposition and subsequent nuisance from blockages or reduced capacity.
Separate system	A drain or sewer system, normally of two pipelines, one carrying wastewater and the other surface water.
Septic tank	A structure for the collection and partial treatment of sewage.
Sewage	Wastewater and/or surface water conveyed by a drain or sewer.
Sewer	A pipe or conduit that carries wastewater or drainage water serving more than one property

Sewer flooding	The unintentional escape of sewage from a sewerage system; the inability of drainage flows to enter a sewerage system because of surcharging.
Sewerage	Alternative term for "drainage collecting system" for foul and surface water systems.
Sewerage system	A network of pipelines and ancillary works that conveys wastewater and/or surface water from drains to a treatment works or other place of disposal.
Sewerage undertaker	An organisation with the legal duty to provide sewerage services in an area. In England and Wales these services are provided by 10 water service companies, in Scotland by three water authorities, and in Northern Ireland by the Water Service of the Department of the Environment for Northern Ireland.
Side weir	A diverting weir constructed on the side of a channel or conduit, usually at right angles to the centre-line of the main channel.
Silt	Sediment (often soil) consisting of particles between 0.002 mm and 0.02 mm in equivalent diameter.
Simulation	The representation of specific conditions during a specific period in a sewerage system, treatment works, river etc, by means of a computer model.
Simulation model	The representation of a physical system and its time-related behaviour by a computer model.
Sluice gate (penstock)	A gate constructed to slide vertically and fastened into or against masonry of dams, tanks, or other structures under which flow takes place when open.
Soakaway	A pit into which surface water is drained to infiltrate into the ground.
Soffit	The top of the inside of a pipe or conduit.
Soil moisture deficit (SMD)	A measure of soil wetness, calculated by the Meteorological Office in the UK, to indicate the capacity of the soil to absorb rainfall.
Source control	The practice of reducing runoff and also pollutants at their source so that they do not enter the drainage system or become significantly delayed and attenuated.
Spill event	A period when an overflow discharges to a watercourse.
Spill frequency	The number of spill events over a given period.
Stilling pond	A small basin into which flow is discharged, which is used to either dissipate energy or trap solids.
Storage	The impounding of water, either in surface or in underground reservoirs.
Storm	An occurrence of a meteorological event, often of rainfall, snow or hail. Used in connection with a phenomenon that is either unusual or of great magnitude, rate or intensity.
Storm tanks	Storage tanks designed to hold most of the stormwater in either sewers or treatment works such that downstream flooding or incomplete treatment respectively is minimised.
Stormwater overflow	A weir, orifice or other device for permitting the discharge from a combined sewer of the flow in excess of that which the sewer is designed to carry.
Sub-catchment	The ground surface area drainage directly to one gully or a collection of gullies.
Surface washoff	The process whereby the rainfall runoff carries surface sediments and dissolved pollutants into the drain or sewer system.
Surface water	Water from precipitation that has not seeped into the ground and is discharged to the drain or sewer system directly from the ground or from exterior building surfaces.

Surface water system	A drain or sewer system that has been designed to carry only surface water.
Suspended solids	Insoluble solids that either float on the surface of, or are in suspension in, water, wastewater or other liquids.
Sustainable drainage	The application of drainage techniques that are considered to be environmentally beneficial, causing minimal or no long-term detrimental impact.
Swale	A grass channel for stormwater collection with shallow side slopes, which is normally dry except during rainfall.
Synthetic rainfall	Rainfall depths or intensities derived from rainfall statistics and not representing an individual real rainstorm.
Synthetic rainfall series	Rainfall time series usually derived by stochastic processes for use in place of a recorded rainfall series.
Tank sewer	A length of sewer with a cross-sectional area in excess of that required for the conveyance of the normal sewer flow, the additional volume being used for the storage of storm sewage.
Time of concentration	Time between the start of a runoff event and the time when the entire catchment is contributing flow to a specific point in the network.
Time of entry	The time taken for surface runoff to reach the entry into the pipe system.
Time series rainfall	A continuous or discontinuous record of individual events generated artificially or selected real historical events that are representative of the rainfall in that area.
Urban drainage	Pipe systems and other related structures to serve an urban environment.
Vacuum sewerage system	A system that operates under negative (sub-atmospheric) pressure to evacuate drainage flows from a property or group of properties; the system may consist of one or more vacuum pumps, a central vacuum reservoir, pipework and interface valves.
Vortex overflow	A type of storm overflow that makes use of the spiralling flow in a vortex to retain polluting material within the pipe system.
Wallingford Procedure	A design and analysis procedure for urban drainage networks (HR Wallingford and Institute of Hydrology, 1981a).
Washoff (of pollutants)	The transport of pollutant mass from the catchment surface during a rainfall event.
Wastewater	Water used and discharged to drain.
Water quality	The chemical, physical and biological characteristics of water with respect to its suitability for a particular purpose.
Water quality standards	Standards set by the national legislation or European Community directives and enforced by regulatory authorities in member states.
Water-table	The surface within soil or rock strata at which groundwater saturation occurs.
Water UK	The organisation representing all water supply companies in UK.
Watercourse	A natural or artificial channel for passage of water.
Weir	An overflow structure across a channel that may be used for controlling upstream surface level, or for measuring discharge, or for both; usually horizontal and constructed as either broad- or sharp-crested.
Wet well	The entry chamber in a pumping station from which water is pumped to a higher level.

ABBREVIATIONS

ADAS	Agricultural Development and Advisory Service
AFD	anti-flooding device
API_5	antecedent precipitation index (over previous five days)
API_{30}	30-day antecedent precipitation index
BMP	best management practices
BRE	Building Research Establishment
BS	British Standard
BS EN	European Standard for use in Britain (see EN)
BWB	British Waterways Board (trading as British Waterways)
C_r	routeing coefficient – used in the Modified Rational Method
C_v	volumetric coefficient – used in the Modified Rational Method
CDM	Construction (Design and Management) Regulations
CEH	Centre for Ecology and Hydrology
CEN	Comité Européen de Normalisation (European Committee for Standardisation)
CIRIA	Construction Industry Research and Information Association
CSO	combined sewer overflow
DEFRA	Department for Environment, Food and Rural Affairs
DETR	Department of the Environment, Transport and the Regions (replaced by ODPM and DEFRA)
DTLR	Department for Transport, Local Government and the Regions
DoE	Department of the Environment (replaced by DETR)
DoE (NI)	Department of the Environment in Northern Ireland
EA	Environment Agency
EN	Europäische Norm (European Standard)
FEH	*Flood estimation handbook* (CEH, 1999)
FSR	*Flood studies report* (IH, 1975)
FSSR	*Flood studies supplementary reports* (IH, 1975–1985)
HA	Highways Agency
HMIP	Her Majesty's Inspectorate of Pollution (replaced by Environment Agency)
HOST	hydrology of soil types
IDB	internal drainage board
IDF	intensity-depth-frequency (relationship)
IF	effective impervious area factor
IH	Institute of Hydrology (replaced by CEH)
MAFF	Ministry of Agriculture, Fisheries and Food (replaced by DEFRA)

$M_5$2day	five-year depth of rainfall in two days
$M_5$60	five-year 60-minute depth of rainfall
NAPI	New Antecedent Precipitation Index
NERC	National Environment Research Centre
NRA	National Rivers Authority (replaced by Environment Agency)
NT	National Trust
ODPM	Office of the Deputy Prime Minister
OFWAT	Office of Water Services (for England and Wales)
PF	porosity fraction (soil storage depth)
PIMP	percentage impermeable proportion of a catchment contributing to runoff – see PR equation in Glossary
PPG	Planning Policy Guidance (DETR; subsequently ODPM, DTLR)
PPG	Pollution Prevention Guidelines (EA)
PR	percentage runoff
pr EN	provisional EN
RP	return period
SAAR	standard average annual rainfall assessed over a period of years
SEPA	Scottish Environment Protection Agency
SOIL	soil type classification used by FSR (IH, 1975) and The Wallingford Procedure (HR Wallingford and IH, 1981a)
SMD	soil moisture deficit
STW	sewage treatment works
SUDS	sustainable urban drainage system
T&CPA	Town and Country Planning Act 1990
T_C	time of concentration
T_E	time of entry
T_p	time to peak (FSR measurement)
TRRL	Transport and Road Research Laboratory (now TRL, Transport Research Laboratory)
TSR	time series rainfall
UCWI	Urban Catchment Wetness Index – describes the wetness of the catchment, usually calculated for the start of a rainfall event
UPM	urban pollution management
WASSP	Wallingford Storm Sewerage Package – the first computer package for the design and analysis of sewage networks
WC	water closet
WRAP	winter rainfall acceptance potential; used by the Wallingford Procedure (HR Wallingford and IH, 1981a)
WSA	Water Services Association
WwTW	wastewater treatment works

1 SCOPE

This guide is intended to assist all those involved with drainage for new developments. It is specifically directed at developments in the UK based on national requirements and international best technical practice. It emphasises the need for a structured approach that integrates the determination of the site layout with landscaping, technical design and Town and Country Planning Act 1990 (T&CPA) consents. The objective is to arrive at the most appropriate drainage system for the site.

Guidance is provided on aspects of drainage design that are not widely available elsewhere. In particular, the guide focuses upon the approval process, sustainable techniques and hydraulic calculations.

Virtually all hydraulic aspects of site drainage are covered in sufficient depth to enable the user to take the correct approach and to understand the principles of design and the criteria used. Although it deals primarily with the hydraulic issues of drainage, aspects such as operation, maintenance and construction are also briefly addressed. The guide understands drainage to include both foul and surface water drainage systems.

Site drainage is moving away from the simple provision of pipes, and the guide details a number of options that UK regulators are keen to promote in the drive towards achieving sustainability. Because of this, the importance of entering into discussion with relevant authorities at an early stage is stressed.

It is intended that this guide will be the general practitioner's first reference point for good practice and advice on sources of more detailed information. This should make it easier to obtain T&CPA approval and facilitate the successful adoption of the drainage network.

The guide aims to provide both the breadth of information needed for site drainage design and the specific details of the procedures that developers should follow to ensure drainage is part of an integrated approach to the development planning of the site.

1.1 Document structure

The guide starts by looking at planning and drainage in general, before considering technical aspects in detail. These technical chapters are ordered in a top-down sequence, dealing first with roofs and going "downhill" to the site outfall. (Construction is generally carried out in the reverse order.) References are provided to allow more detailed information to be obtained from specialist drainage guides and elsewhere.

The appendices provide additional information on subjects ranging from the roles and responsibilities of regulatory bodies to principles relating to vacuum sewerage. Worked examples are also provided to illustrate the method of approach for designing roof drainage and attenuation storage structures.

A glossary of terms and abbreviations is provided on page 7 to define the meaning of technical words and phrases.

2 THE T&CPA PROCESS

All projects pass through several stages of design before construction proceeds. Various drainage-related aspects need to be addressed in each phase. Very small developments will not necessarily have to consider all of the issues that need to be addressed by large developments, but they should be specifically omitted rather than merely forgotten. Figure 2.1 illustrates the phased approach and the issues to be considered. Figure 2.3 is a chart produced by the Royal Institute of British Architects (RIBA) that also addresses the construction stage and the CDM aspects of health and safety.

At each of the project phases the following three aspects need to be specifically considered. These are:

- technical issues
- duration of each phase
- T&CPA approval activities.

2.1 T&CPA considerations for the developer

From the area structure plan and local plans produced by county councils and district councils, the developer can see the land that has been allocated for development and the land use category applied. The problem facing developers is that little public information is available for Town and Country Planning Act (T&CPA) guidance on the drainage requirements/limitations for sites. Experienced developers know that it is important to establish what drainage options exist before putting in a T&CPA application, as the cost involved in meeting criteria set by the relevant authority is often considerable.

The Environment Agency provides general guidance in the form of Local Environment Agency Plans (LEAPs), but detailed negotiations are likely to be needed for individual sites. The Agency plays an important role in providing guidance to local authorities for proposed developments.

The option of draining properties to a public sewer, if it exists, can always be taken, as the sewerage undertaker has a legal duty to provide such a service. The receiving capacity of the sewer is often limited, however, and on-site storage is often stipulated. This aspect can only be established by discussion with the sewerage undertaker. Although the highway authority is responsible for road drainage, developments rarely build road drainage separately from pipes serving roof drainage, because this results in a three-pipe system (foul, roof and road) serving the site. Thus surface drains tend to serve both roads and roofs. The main exception to this is where roofs and/or roads are served by soakaways.

If there is no sewer locally, a sewer can be requisitioned. A developer who requests such a sewer has to pay an annual charge for the provision of the sewer for up to 12 years, though a single commuted sum is often agreed to. The basis of this charge is defined under Section 98 of the Water Industry Act, 1991.

The developer is not only interested in securing T&CPA approval for surface water disposal, but is usually concerned to have the sewers "adopted" by the sewerage undertaker or, in some cases, the highway authority. This imposes certain standards of construction, and also constrains the developer to using drainage options that are acceptable to the relevant authority. The authority is usually concerned to minimise long-term maintenance costs, so it is often reluctant to consider options other than a traditional pipe system. This aspect is particularly of concern to the Environment Agency and similar bodies that are trying to apply more environmentally beneficial solutions to site drainage. Sustainable drainage systems are detailed in the National SUDS Working Group's *Interim code of practice for SUDS* (NSWG, 2004). In addition, design guidance is provided in CIRIA publications C521, C522 (Martin *et al*, 2000a and b) and C609 (Wilson *et al*, 2004).

The T&CPA process is illustrated in Figure 2.2.

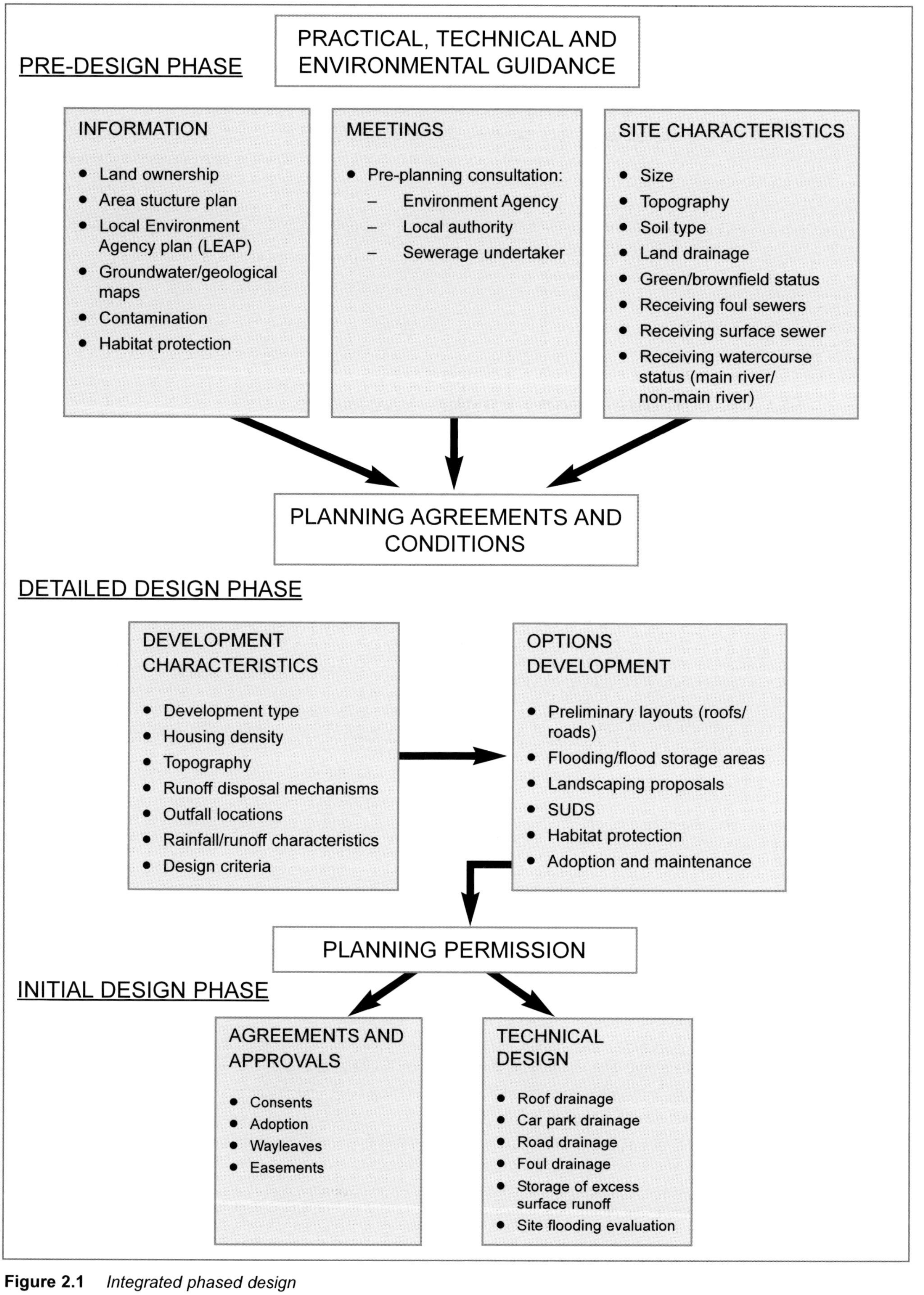

Figure 2.1 *Integrated phased design*

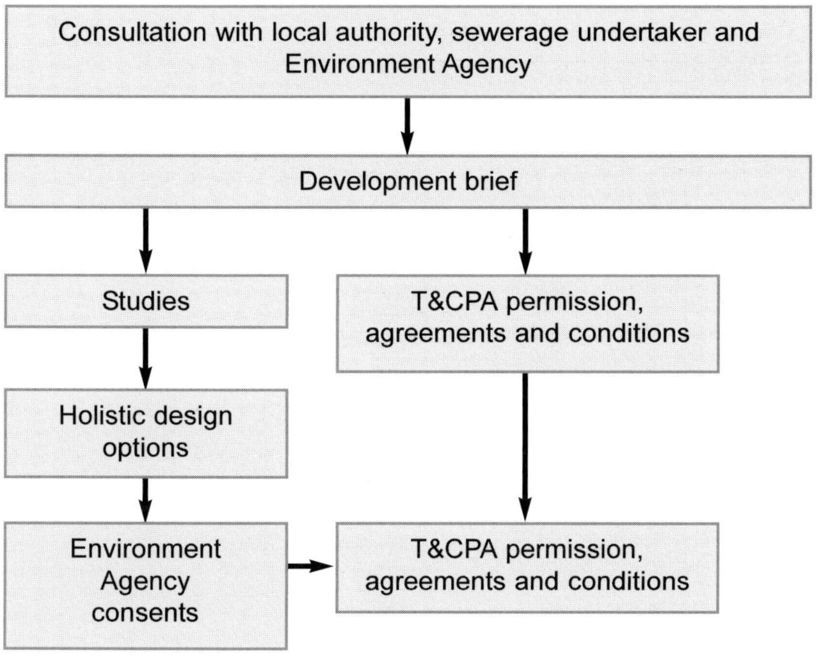

Figure 2.2 *The T&CPA process*

2.2 Stages of phased management of the development

Design is generally an iterative process that starts from an initial concept and is gradually refined to produce an optimum solution. The final solution should provide best value for money against the project objectives within political, economic, environmental, social and technical constraints.

To manage the design process effectively, the design should progress though a series of steps, with increasing levels of detail as it moves forward. At the end of each step, the solution should be evaluated against the project objectives and formally agreed with the client before proceeding to the next step.

The process is shown in Figure 2.3, which should be modified to suit the requirements of individual clients and projects.

It should be noted that the degree of cost certainty increases as the project progresses, ranging from only about ± 30 per cent cost certainty at the end of feasibility to about ± 5 per cent at the end of detailed design.

RIBA stage	Brief description	Main actions	Key outputs
1	**Feasibility**	• Confirm development brief • Make initial contact with EA/SEPA, sewerage undertaker, local authority, other services highway authority • Obtain ground conditions • Identify site constraints • Investigate disposal options • Identify flood paths • CDM risk analysis	• Preferred point of foul disposal • Preferred means of surface water disposal • Abnormal features/costs • Requirements for additional surveys • Flood potential • Identify principal health and safety hazards
2	**Outline design** Need for sub soil drainage Any drainage diversion Established design parameters CDM risk analysis Key levels	• Layout of principal foul and surface water drains in plan • Layout of principal foul and surface water drains in long section • Estimate discharge rates • Calculate sizes of drainage features (eg pipes, separators, balancing) using approximate methods • Method of paved drainage (ie gullies, linear drains etc)	• Agreement in principle of adoption • All principal elements, eg separators, pipework layout, balancing, for costing • Input into health and safety plan
3	**Detailed design**	• Confirm all foul connection points • Confirm all surface water connections • Access arrangements (ie manholes, inspection chambers) positions, levels and sizes • Select pipe materials and calculate strengths and bedding • CDM risk analysis • Check for pipe clashes	• Position level and size of all pipework and chambers • Drawings and specifications for tendering and legal arrangements • Input into health and safety plan
4	**Production information**	• Prepare details of structures • Prepare maintenance guidance	• Sufficient information for construction and maintenance • Input into health and safety file

Figure 2.3 *Design process*

2.3 The T&CPA procedure

Obtaining T&CPA approval and discharge consent for site development drainage can cause considerable frustration. The process is quite complicated, partly due to the number of regulators and authorities that are involved. This section provides guidance on the approach needed to obtain consent requirements by the organisations involved, and the procedures and requirements of the T&CPA process.

The developer has to go through a process of submitting site development proposals to the local authority to get approval to implement them. This procedure has several stages between the initial application and the start of construction, which is illustrated in Figure 2.2.

Preliminary pre-application consultation is often beneficial and avoids unexpected problems and delays at the T&CPA application stage. Time should be allowed for option reviews and collection of data.

Although the local authority is responsible for this process, it normally involves other organisations such as the Environment Agency or the sewerage undertaker in obtaining approval for drainage proposals.

The principal regulatory authorities involved are summarised in Table 2.1.

Table 2.1 *Principal regulators and their responsibilities*

Country	Regulator/authority	Responsibility
England and Wales	Environment Agency	Groundwater and controlled waters Main river consents Non-main river advice/consents
	Internal drainage board	Non-main river consents
	Local authority	T&CPA approval Non-main river consents [1] Building Regulations [2] Highway drainage
	Sewerage undertaker	Sewer consents Adoption of sewers
	British Waterways Board, National Trust etc	Discharge consents
Scotland	Water authority	Sewer consents
	Regional and city councils	Flood defence River discharge consents
	SEPA	River discharge consents (water quality) Groundwaterconsents
Northern Ireland	DoE (NI)	Sewer consents Groundwater consents River discharge consents

Notes

1 The local authority generally receives advice from the Environment Agency (statutory consultee) before issuing consents for non-main rivers

2 Or other approved building control organisation.

Figure 2.4 provides a simple overview of the responsibilities of organisations in the UK that might be involved in dealing with drainage aspects of a development.

A more detailed description covering the interaction of all the possible bodies involved is given in CIRIA Report 124 (Maskell *et al*, 1992).

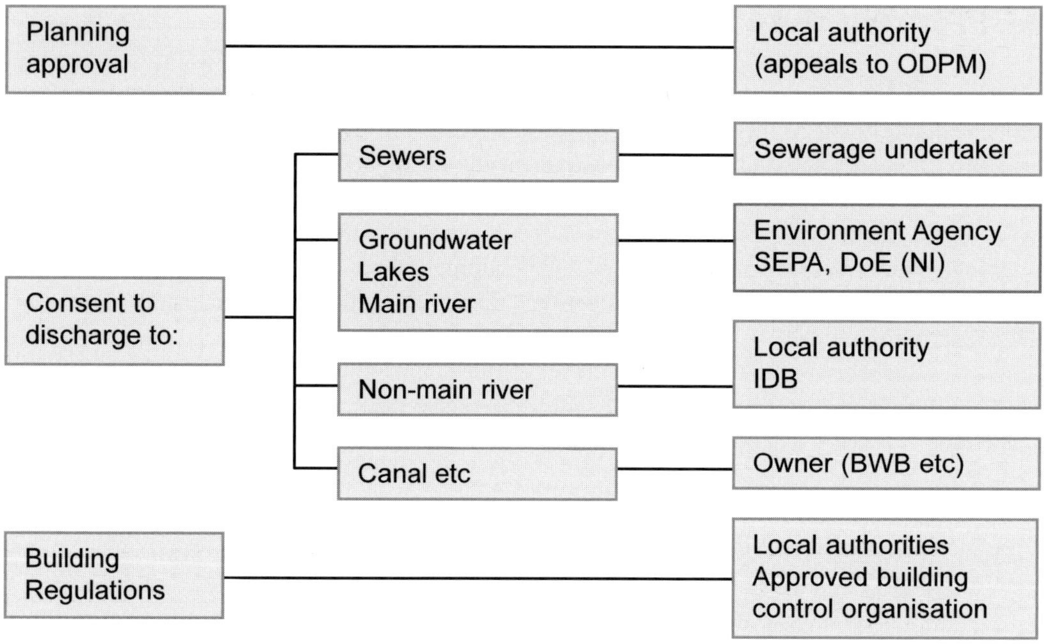

Figure 2.4 *Responsibilities for T&CPA, discharge consents and Building Regulations*

The T&CPA application form asks the question: "How will surface water drainage be disposed of?". The options, or combination of options, available to answer these questions are:

- soakaway or other on-site infiltration
- public sewer
- highway drains
- non-main river
- main river.

After any pre-application consultation, the process for considering drainage proposals is as follows.

1. A T&CPA liaison officer in the local authority reviews the application to assess whether the drainage proposals require a formal meeting between the authority, the Environment Agency, the sewerage undertaker or other bodies.
2. The local authority (usually in conjunction with the Environment Agency) considers:
    - the risk of flooding on the site (Environment Agency floodplain policy)
    - environmental aspects of the site (including habitat protection)
    - opportunities for sustainable drainage techniques
    - geological characteristics of the site
    - the need for on-site attenuation (of surface water runoff)
    - the limitations of off-site drainage capacity.
3. The drainage officer within the local authority advises on T&CPA consultation with the Environment Agency (a statutory consultee under the Town & Country Planning Order 1995) based on the categories listed in *Liaison with local planning authorities* (Environment Agency, 1997a). The Environment Agency may advise the developer to review the requirements for the site.
4. The sewerage undertaker is consulted on larger developments where public sewerage systems are to be connected. Both sewerage capacity and sewage treatment are considered. Water supply is also assessed, but lack of capacity does not necessarily lead to a refusal of T&CPA approval.
5. Legal agreements may be needed between the developer and the T&CPA authority.

2.4 Categories of drainage design

This section provides a summary of the drainage design activities and the authorities that are involved. Each of these aspects is considered in more depth in Chapters 4 to 10 of this guide.

Table 2.2 summarises the various aspects of site drainage, the authority involved in the approval process and the key documents that define criteria and the requirements or standards that need to be met. Some criteria are location-specific, so they are not provided in the reference documents. In these situations, discussion with the authority responsible is needed to determine the requirements. This is particularly true for control and storage of runoff.

The table refers to authorities in England and Wales. The equivalent organisation (as detailed in Section 2.3) should be referred to in Scotland or Northern Ireland. Appendix A provides more detailed information on the authorities and their roles.

Table 2.2 *Design responsibilities and key documents in the UK*

Drainage issue	Responsible authority	Key reference documents
Runoff rate to watercourse (post-development)	Environment Agency Local authority	The Wallingford Procedure (HR Wallingford and IH, 1981a)
Surface runoff into drainage system (post-development)	Sewerage undertaker	*Sewers for adoption* (WRc, 2001) CIRIA Report 124 (Maskell *et al*, 1992)
Design of sewers	Sewerage undertaker	*Sewers for adoption* (WRc, 2001)
Protection of groundwater	Environment Agency SEPA	*Policy and practice for the protection of groundwater* (EA, 1998a)
Building Regulations	Local authority	Building Regulations (England & Wales) Part H, edition 2002
Roof drainage	Local authority	BS EN 12056-3:2000
Road drainage Car park drainage	Local authority	BS EN 752:1997 BS EN 1433:2002 HA 37/97 (Highways Agency, 1997)
Flood risk	Environment Agency	PPG25 (DTLR, 2001)
Soakaways	Environment Agency Local authority	CIRIA Report 156 (Bettess, 1996) BRE 365 (BRE, 1991)
Floodplain development	Environment Agency	PPG25 (DTLR, 2001)
Runoff from greenfield sites	Environment Agency Local authority	FSSR 16 (IH, 1985) MAFF Report 345 (MAFF, 1981) Report 124 (IH, 1994) HR Report SR 591 (Kellagher, 2002b)
Sustainable drainage systems	Environment Agency SEPA Local authority	CIRIA C521 (Martin *et al*, 2001a) CIRIA C522 (Martin *et al*, 2001b) CIRIA C523 (Martin *et al*, 2002) CIRIA C609 (Wilson *et al*, 2004) *Interim code of practice for SUDS* (NSWG, 2004)

Section 2.6 provides details of all these key reference documents.

2.5 Building Regulations

Different legislation applies to building works in England, Wales, Scotland and Northern Ireland, as shown in Table 2.3.

Table 2.3 *Building Regulations legislation in the UK*

England and Wales	Scotland	Northern Ireland
The Building Regulations 2000	Building Standards (Scotland) Regulations 2001 – amendment to 1990 Regulations	Building Regulations (Northern Ireland) Order 1997 – amendment to 1994 Regulations

Drainage works are normally covered by Building Regulations, although there are exceptions such as agricultural activities. These Regulations are written to meet the requirements of the Building Act 1989. The Building Regulations in England and Wales require drainage to "be adequate". Guidance on what is considered to be adequate is given in Approved Document H to the Building Regulations, edition 2002, which, as well as giving advice for domestic dwellings, also refers to British Standards on the subject. For works in and around buildings, Building Regulations approval is needed. The "alternative approach" to satisfying the requirement for drainage to be adequate is to comply with recognised standards

The Building Regulations also cover sanitary conveniences and washing facilities. The main matter relating to measures for the prevention of sewer flooding is the requirement in Approved Document G1 to the English and Welsh Building Regulations, which stipulates that householders must have access to a WC connected directly to a gravity drainage system.

The latest revision of Building Regulations Part H (edition 2002) aims to bring about a convergence in the standards of construction of private and public sewers.

Building regulation control is exercised by local authorities and in certain instances also by approved independent companies.

3 SITE LAYOUT DESIGN

Site drainage options and layout can often be influenced or even dictated by conditions found at the site. It is important to consider drainage within the initial design layout of the site to ensure that best use can be made of existing conditions. This chapter highlights the areas that should be considered, including:

- drainage philosophy
- flat sites
- steep sites
- disposal of surface water
- disposal of foul water
- soil type
- flood storage
- floodplains and river corridors
- industrial and commercial sites
- landscaping
- environmental issues
- contaminated land
- size of development
- water reuse
- land drainage
- foul/roof/road pipe systems
- vacuum sewerage
- strategic installations
- areas of risk particular to site drainage
- detailed design issues.

3.1 Drainage philosophy

This section considers the aims and objectives used in the design of site drainage. A new site to be developed is often assessed at outline design in terms of maximising its potential for commercial return. The tendency is thus towards a high density of development, which results in the generation of large surface runoff flow rates and volumes. This is contrary to the emerging philosophy of sustainability, which aims to achieve post-development runoff behaviour similar to the greenfield response of the site. Sustainable urban drainage systems (SUDS) have been developed to try to meet this objective. This debate has been sharpened with the perceived increase in flooding incidents and the research findings from climate change studies.

To help achieve a "sustainable" drainage system, a planner or architect needs to assess the site's drainage requirements before the site layout has been defined. It is important to be aware of the issues that will alleviate problems at a later stage when the drainage system is being designed. The following general principles need to be considered.

1. The topography of the site should be used to ensure that sewers, pumping stations and pumping mains can be laid in roads and not in private property and stormwater storage is appropriately located.
2. Design parameters and criteria.
3. Whatever the design criteria used, extreme rainfall events can take place at any time.
4. The locations of the outfall points for both foul and surface water drainage should be clearly established at a very early stage.

5 Any need for site storage for rainfall runoff should be agreed together with criteria relating to the constraints to be applied to outfall discharges.

6 Phased developments require the total site development to be known to ensure pipes and tanks are sized appropriately.

7 Sewers, pumping stations and pumping mains to be adopted are laid in publicly accessible places.

8 The use of SUDS techniques should be investigated to assist in complying with the concept of sustainability.

9 Runoff control and flood routeing is linked to landscaping and site layout.

10 Develop an understanding of ground conditions; presence of rock, high groundwater, old mine workings, quarrying, contaminated soils.

3.1.1 The layout of proposed site development

<u>Access to sewers for maintenance</u>
Roads, car parks and open areas serving a development site should be designed so that drainage pipework can be laid under them to allow access for maintenance. If this aspect has not been considered in the initial site layout design together with the general topography, the sewer network system may have to pass across private property, or be laid at great depth for some sections or, in extreme cases, require pumping stations. If adoption of the sewer system is expected, it is usually necessary to ensure that access does not entail entry to private land.

<u>Use of sustainable urban drainage systems (SUDS) within landscaping proposals</u>
The Environment Agency and other authorities are keen to promote the use of sustainable practices. The application of SUDS for drainage means that careful attention needs to be given at an early stage to site layout and landscaping. In some areas, the term "best management practices" (BMPs) is used to mean much the same thing, implying that best practice involves the use of SUDS techniques.

<u>Commuted sums for maintenance of structures</u>
If pumping stations are built, commuted sums may have to be paid to meet their operation and maintenance costs. Although this is rarely the case in practice, sewerage undertakers try to avoid the use of pumping stations where possible. Similarly, if SUDS systems are used and need regular maintenance, the adopting authority might require a commuted sum.

3.1.2 Extreme events

<u>Overland flood flow risk</u>
Surface water drainage is normally designed not to surcharge during rainfall events with a return period of one or two years and generally has about a 1 in 30-year "no flooding" service criteria. However, more extreme events do take place, causing significant overland surface flows and flooding at low points. On steep sites it is important to assess overland flooding, and property should not be located in the likely path of flooding.

<u>Downstream constraints to discharge</u>
Where there are downstream constraints to discharge, planned storage areas should be identified for on-site retention.

<u>Property floor levels</u>
On flatter sites and floodplains where extreme events can cause flooding and ponding, the location of key services such as telephone boxes and transformers, as well as the more obvious issues of property location and floor levels should all be considered. Floor levels should normally be at least 150 mm higher than the maximum flood level that could take place in an event of a 100-year return period. Greater freeboard may be appropriate under certain circumstances, particularly where records exist of flood levels that are higher than the estimated 100-year level. The local authority or the Environment Agency should be able to advise on likely flood levels.

Roads acting as flood channels
Roads and car parks should be specifically utilised to store or channel extreme surface water runoff to defined storage areas or streams. This is often referred to as major and minor drainage and is formally practised in countries such as Australia. It should be remembered that water can come out of sewer systems as well as enter them depending on the hydraulic gradient relative to ground levels. Specific consideration should be given to the depth and frequency of flooding to ensure that the level of service is adequate.

Flows into and from the site
There is a legal obligation to prevent discharge of surface water from the site into neighbouring areas and also to prevent natural flows from entering the site and being picked up by the surface water drainage.

Flood return periods
Development in areas liable to flood is covered in Sections 3.7 and 3.8, but it should be noted that 1 in 100-year return period is a general standard that is applied to catchment flood risk. Alternatively, the worst historic recorded flood level is often used. If the site is below a reservoir, considerably return periods need to be used – reference should be made to the requirements of the Reservoirs Act 1975.

Recent events in the UK and Europe have prompted debate about flooding, and the insurance industry is now making reference to the 500-year return period, while the Environment Agency is using a return period of 200 years in some coastal regions. A further debate is taking place over the differences between the FEH (CEH, 1999) and the FSR (IH, 1975) and the anticipated impact of climate change. It is therefore important to agree on the level of risk to be considered and the tools to be used in determining potential flood impact.

Special needs for development in tidal areas and river corridors
Developments in areas protected from tidal flooding need special consideration. Procedures should be prepared defining actions and checks on drainage installations. Return periods of 200 years are regularly stipulated in these circumstances.

3.1.3 Outfall location points for foul and surface water drainage

Receiving manhole or stream for surface water discharge
The outfall location points for both foul and surface water needs to be established. The location of the receiving manhole or stream should be determined together with information on the invert level and pipe size within it. Similarly, infiltration points should be located with due regard to foundations and preferential ground conditions.

Trade waste charges
The sewerage undertaker will assess the foul sewer connection with respect to the quantity and effluent characteristics of the sewage to establish that the development can be served. A financial contribution may be required to uprate the receiving network due to the increased load. Financial charges are sometimes levied for treatment of trade wastes if suitable on-site pre-treatment is not carried out.

Discharge constraints
Chapter 2 highlighted the range of organisations that might be involved in giving approval to a surface water discharge. Constraints are often placed on the discharge rates from the site that require the developer to use storage. This storage can range from a throttle downstream of over-sized pipes to a range of attenuation or interception systems. If the soil conditions are suitable, infiltration solutions can be used to reduce the flows from the site. Chapters 8 and 10 discuss a range of attenuation or flow reduction techniques that are available.

3.1.4 Surface water storage

Tank sewers in roads
Surface water storage may be achieved using ponds, tanks or tank sewers. If storage is used, this often introduces problems of fitting the system on to the site. Both flat and steep catchments have constraints related to levels, location

and access. Tank sewers are difficult to place in steep catchments if roads have not followed the contours of the site. In flat sites it is often difficult to use tanks and to achieve a high enough outfall level to connect to the receiving sewer or river.

Risk of blockage for control orifices

Tank sewers are often used with criteria such as a discharge limit of 5 l/s/ha. Values can range from 1 l/s/ha to 20 l/s/ha. This often leads to the use of very small control orifices for smaller development sites, which increases the risk of blockage. A minimum orifice size of 150 mm is generally specified. Overflow structures and maintenance issues must always be considered. Installing a light liquid separator upstream of a throttle offers a degree of protection against blockage and minimises the risk of pollution. Vortex-type controls are less prone to blockage than orifices or throttle pipes, as orifices are larger for any given capacity.

Sewerage undertakers may take different views on the most appropriate form of storage (tank sewers or ponds) for attenuation purposes, but a number of issues have to be analysed, particularly operation and long-term maintenance implications, before these attenuation structures are accepted.

Detention or retention ponds for flat sites

It may be more appropriate to use ponds for flat sites for several reasons. First, the problem of providing minimum cover to the storage structure is not needed. Second, it avoids the potential problem of flotation when stormwater storage tanks are empty in areas where the groundwater table is high. The design of the pond should take account of the groundwater level and its seasonal fluctuations, particularly if liners are to be used.

3.1.5 Foul sewage storage

Foul sewage storage at pumping stations needs to be provided in many instances where the implication of pump failure makes it a requirement. Often the criterion is for three hours of storage at 2 × DWF. Telemetry linkage to the operations section of the sewerage undertaker then allows an emergency response to take place before sewage flooding occurs.

In special circumstances where an industrial unit (such as a hospital) discharges effluent with very highly varying flow rates, storage may be needed to limit peak flows from entering the receiving sewer. Similarly, storage may be required to allow hot discharges (from a laundry, for example) to cool.

3.1.6 Phased developments

Sewer sizing and velocity assessment for all phases

The first phase of a multi-phased development can be located either at the upstream end of the catchment or the downstream end. However, when sizing and locating the principal trunk sewers, a knowledge is needed of the future site development. The sewers should be designed to operate with both the reduced flows of the first phase, but also sized large enough to serve subsequent runoff from the other phases.

The Old fixed PR Wallingford Procedure equation

The Wallingford Procedure runoff model is a correlated equation based upon normal fully urbanised areas. This runoff model can be misused in situations where large permeable areas for future development are included in the analysis of the first phase of development. Care should be taken in applying the equation appropriately.

3.1.7 Sustainable drainage techniques

Chapter 8 details sustainable drainage techniques that are available for site drainage. Various techniques are being recommended to limit runoff from a site. Some can be difficult to apply due to the limited experience in their use and uncertainty regarding their long-term performance. Sewerage undertakers are therefore reluctant to adopt certain systems and have particular concerns about the maintenance costs and the long-term implication of potential unplanned connections to the sewer system. Their environmental benefits are widely acknowledged, however, and the Environment Agency strongly recommends their use where appropriate.

3.2 Flat sites

Specific issues affecting flat sites include:

- drowned outfalls
- minimum sewer depths
- velocities/sedimentation
- infiltration techniques
- storage tanks
- flushing siphons
- flooding and ground saturation.

3.2.1 Drowned outfalls

Outfalls, either to a river or a sewer, in flat catchments often have to operate with the downstream water level surcharged above the outlet pipe level. Care should be taken in the hydraulic design of pipe flows either to take the water level into account or to have a reasoned argument for ignoring it. Similarly, if pond or pipe storage is being used with a top water level above the incoming sewer, the hydraulic implications should be considered in the pipe design.

Pipe design is often carried out using the Rational Method, but it is difficult to use this technique if a downstream water level is above the invert level of the outfall pipe. Where this is the case it is advised that analysis be carried out using a hydrograph simulation method. For small sites, a flat-rate rainfall intensity may be used in conjunction with hydraulic gradients (taking into account the downstream water level), which enables experienced drainage engineers to define a draft outline network very rapidly.

<u>Combined probability</u>
If discharge is taking place into a river or large receiving sewer, establishing the design requirements for a specific level of service is not straightforward. The level in a stream for a given return period may be known, but using that value in the analysis can result in a very conservative assessment. This is because the critical storm duration for a small site is likely to be less than an hour (or a few hours if storage is needed), but the critical duration for the river may be a few hours up to two or three days. In addition, the achievement of a specific level of service should theoretically check a range of combinations of rainfall and downstream constraints to establish the worst-case event. It is theoretically possible to determine the probability of each combination (although the complexities of dependency are not discussed here), but a degree of pragmatism is usually applied by making some simple, though conservative, assumptions.

3.2.2 Minimum sewer depths

<u>Minimum depth</u>
Flat sites are often difficult to drain due to the constraint of levels of existing outfalls. Pipes need to be laid at gradients to achieve minimum velocities (referred to as "self-cleansing velocity") to keep the pipe clear. To minimise the depth of the sewers it is common practice to reduce the cover over pipes at the upstream end of runs. Chapter 9 on pipe design drainage provides some information on possible minimum pipe depths and bedding requirements. The traditional design limits are 900 mm cover under roads and 600 mm at other locations. The use of drainage channels instead of pipes provides a means of maximising available falls.

<u>Backwater effects</u>

In circumstances where a tank sewer is required (see Section G4), there is sometimes a temptation to build it with equal inverts rather than equal soffits where the small site sewer enters the tank sewer. This might be technically acceptable so long as consideration is given to sedimentation risk and hydraulic backwater effects in the incoming pipework. There is usually a positive fall along the length of the tank with a gradient of 1 in 100.

3.2.3 Velocities/sedimentation

Minimum velocity

There are two parameters that can be modified to minimise pipe depths on flat sites. The first is to reduce the minimum pipe cover, and the second is to reduce the velocity. Minimum velocity guidelines exist, but it is possible to use flatter gradients if more detailed analysis is carried out on expected flow rates and sediment transport characteristics, provided workmanship can be guaranteed and ground conditions are such that settlement is unlikely.

Egg-shaped pipes

Egg-shaped sewers are not commonly used for modern sewers in the UK, and although they are available, they cost more than circular pipes. However, the hydraulic benefits of the flatter gradients to which ovoid pipes can be laid can also result in a more economic solution in some circumstances.

3.2.4 Infiltration techniques

Use of soakaways

Infiltration is dealt with in more detail in Appendix G. Infiltration is a good method of reducing site discharges, especially as onerous discharge limits are often applied to flat sites. Care must be taken to ensure that the soil characteristics of the site are suitable and that the solution remains viable with winter groundwater levels.

Groundwater-sensitive zones

Discussions must be held with the regulator to make sure that the site can use soakaways. Infiltration may not be allowed due to potential groundwater contamination or the area may have been zoned as a sensitive groundwater location.

3.2.5 Storage tanks

Storage on flat sites

The main constraints to designing storage tanks or attenuation ponds on flat sites are:

- an outflow high enough to drain by gravity to the receiving sewer or stream. If a pond is a wet one then this is not usually a constraint. Pumping is considered where gravitational drainage is not possible
- the maximum water level does not hydraulically limit the effectiveness of the incoming sewers
- the shape of the structure may be constrained by minimum cover if a tank sewer is being used
- a high risk of sediment deposition, and increased maintenance demands, caused by the drain-down rate and the velocity characteristics in filling and emptying the tank sewer
- the risk of tank flotation in water-logged ground when the structure is empty.

An obvious point, but worth stating, is that it is normal to locate the pond or tank at the low point of the site. This needs to be taken into account early in the process when proposals for layout and landscaping are being conceived.

Excess floodwater storage

An alternative to designing a large storage unit is to build a smaller one to cater for the frequent events and allow rare events to flood in designed locations such as car parks or public open space. This has the advantage of minimising the cost of the storage structure, minimising the land needed and reducing the size and therefore also the depth of the storage structure. The resulting repercussions of occasional flooding and potential for complaints should be considered when utilising temporary surface flooding.

3.2.6 Flushing siphons

Flushing siphons to keep foul drains clear have been used in many parts of the world to provide a flood wave to move solids down the sewer. Where this is considered due to housing density or low water consumption, it is preferable for the siphons to be served by greywater (bath and hand-basin wastewater) and not all domestic wastewater.

3.2.7 Flooding and ground saturation

Flooding from overland runoff from other parts of the site is not likely to be a problem. Flooding is more related in these situations to ground saturation and rising groundwater levels. This can be serious if infiltration is the preferred drainage mechanism, and in certain instances results in weeks of saturated conditions.

3.3 Steep sites

Steep sites have their own particular problems that need to be addressed and these include:

- extreme event flood risk
- pipe gradients
- storage tanks.

3.3.1 Extreme event flood risk

Overland flood flow paths and location of properties
If sites are steep, specific consideration should be given to flood flow, looking at both steep slopes and road layout. Overland flood flow (outside of floodplains) may not seem to be a real issue in UK, but occurrences take place sufficiently frequently to justify making a point of considering this problem. Sites on clay or chalk can suffer from flash flood events with rainfall runoff approaching 75–100 per cent. Instances have been known of sheet flow over long grass and on sites where the contractor has stripped topsoil prior to development. In one instance, the first phase of a development had been located at the downhill end of the site and mud was washed through new houses in three flood events over an 18-month period. Properties should not be located where significant overland flow would occur unless some form of channelling is provided.

Flows into and from the site
There is a legal obligation to prevent discharge of surface water from the site into neighbouring areas and also to prevent natural flows from entering the site and being picked up by the surface water drainage. Although this is of general application it is likely to be more of an issue in steeper catchments.

Alignment of roads
Roads on steep catchments can act as channels. Water can surcharge out of a sewer and flood out on to a road. If roads are laid down the hill rather than along the ground contours, the result can be a torrent that discharges at the bottom of the road to whatever is in its path. Standard British gully design is unlikely to cope with this type of circumstance.

The use of the Rational Method does not allow the analysis of the potential impact of surcharged flows and flooding.

3.3.2 Pipe gradients

Pipe capacity at high velocities
Traditionally, pipes were laid at gradients to keep the flow velocity within an envelope of a minimum and maximum velocity range, with the maximum set at 3.0 m/s. This limit was thought to be needed to minimise scour. However, although erosion is an issue, particularly at around 3.0 m/s, relatively recent study indicates that turbulence at higher flow rates actually reduces this problem. The main limitation of high velocities is the bulking caused by air entrainment, which reduces the actual capacity of the sewer by up to 20–30 per cent when running in surcharge; allowance should therefore be made for this. The problem of cavitation and the related structural effects at very high flows only starts where velocities approach 20 m/s. In practice, therefore, there is no restriction on how steep a pipe may be laid, although pipes may need to be anchored to prevent slippage.

Discharge velocities to open streams

For pipes discharging to open streams, baffles and alignment as well as bank protection should be such that erosion of the bank does not take place.

Safety issues in large pipes with high velocities
It is important to remember the safety of personnel entering sewers with high velocities and flow rates, in addition to the normal precautions necessary before entry to any confined space.

3.3.3 Storage tanks

Storage tank upstream soffit level relative to downstream ground level
Storage tanks are also an issue for steep catchments. A common form of storage is the use of tank sewers (see Appendix G). These tend to be relatively long, often using 900 mm or 1200 mm pipes. If the catchment is steep the alignment of the tank needs to take into account that the upstream soffit must be below the cover level or the emergency overflow level of the tank to mobilise all the storage. This often means that the alignment should not be in a road that is coming down a hill as the upstream end may need to be constructed at depths of 5 m or more. It is better to try to align the tank with the site contours.

3.4 Disposal of surface water

Surface water can be discharged from a site in three ways: infiltration, discharge to river and discharge to sewer system.

Infiltration and discharge to a river
As a statutory consultee, the Environment Agency is always involved in discussions relating to infiltration to ground and discharge to a river – even one not designated as a main river. Although the principle of groundwater recharge is encouraged, the risk of groundwater contamination is always an issue in discussions related to the Town and Country Planning Act 1990 (T&CPA).

Discharge to a sewer or drainage system
Sewerage undertakers are involved in accepting flows from sites, whether foul or surface runoff. Where existing pipe systems cannot provide an acceptable level of service, the developer will often have to limit the flow rate from the site to ensure the system downstream can cope with the flows. In some circumstances developers may need to model the potential impact or even improve the receiving sewer system. Not all drainage systems are the responsibility of sewerage undertakers. The local authority is the highway authority and is also responsible for land drainage, so it is important to establish who owns the local drainage system to which connection is envisaged.

3.5 Disposal of foul water

Foul water is normally drained to a public sewer and adopted by the sewerage undertaker. Occasionally developments take place in remote locations and wastewater treatment works need to be used. Wastewater treatment can be either package systems or designed for treatment using sustainable treatment methods. Traditionally, septic tanks (CIRIA SP144BT, 1998 and BS 6297:1983) have been used in these circumstances and are still often considered. Research has found that the infiltration of the effluent needs to take account of the biofilm build-up in the soil, making it relatively impervious. Sizing of infiltration units cannot therefore be based on standard soakaway criteria.

Where local treatment is being considered it is always necessary to discuss the proposal with the Environment Agency and to obtain the appropriate consent to discharge.

3.6 Soil type

The soil type of the development site is a significant issue both fduring construction and for drainage system design.

Volume of runoff
It is important to be aware that soil type affects both the drainage options available and the amount of rainfall runoff. Sites composed of sandy soils suffer from a proportionally greater increase in runoff after development than clay sites. This has implications with regards to applying the philosophy of no change compared to the pre-development state, with greater consideration needed for reducing volumes and flow rates discharged from the site.

The use of infiltration

Sites that have rock under the topsoil have high levels of runoff and do not allow the option of infiltration unless the rock is well fractured. Clays have similar rainfall runoff characteristics. Sand and gravel soils have lower runoff and allow greater freedom in considering drainage options.

Temporary sediment storage ponds during construction

Suspended solids in runoff are a particular problem during construction. Precautions are usually required to ensure that all rainfall runoff passes through temporary storage ponds to settle out most of the sediment.

3.7 Flood storage

Extreme events can lead to flooding. Flood storage design falls into two main categories (Hall *et al*, 1993):

- temporary storage
- long-term storage.

3.7.1 Temporary storage

Extreme event consideration of flooding

Criteria for drainage design should consider more than just pipe performance. The pipe system will be designed for a level of service defined by the sewerage undertaker, but it will be necessary to demonstrate and cater for the behaviour of potential flooding for higher return periods. It is rarely necessary to model the overland flooding on a site to predict flow rates and volumes, but it is essential to understand the consequences of flooding taking place at any location. The designer should therefore avoid putting structures in the flood regions and should consider planned temporary flooding areas (see Section 3.7.2), which may be car parks or recreational areas (see Section 10.2.3).

3.7.2 Long-term storage

Volumetric compliance to sustainable drainage

The phrase "long-term" is used only to distinguish it from "temporary" (see Sections 8.2.2 and 10.2.3) to emphasise the need to store water until the flood risk has passed. Temporary storage on site, whether for extreme events or as part of the system design to limit flows to the river, results in all the water passing to the river relatively quickly. Research by HR Wallingford (Kellagher, 2002a and b) has shown that temporary storage is generally not very effective in protecting the river during periods of flooding from the effects of increased runoff caused by developments. An alternative strategy has developed whereby direct drainage to "long-term" on-site storage is specifically not provided. Soakaways comply with this concept. An alternative approach is to include the designed flooding of areas that would be drained by sub-surface land drains and so retain the floodwater for several days. This ensures that the volumetric runoff from a site is not increased compared with the pre-development state. If using this concept, the designer must plan and design runoff storage and set out land drainage methods that ensure the area is drained appropriately, considering rates of runoff and infiltration of the various techniques.

3.8 Floodplains and river corridors

3.8.1 Floor levels in floodplain development

The Environment Agency normally objects to developments in floodplains. This is due to the need to preserve flood flow and flood attenuation characteristics for extreme event river flows. If development is allowed, however, special consideration needs to be given to property floor levels. The normal criterion is to have floors set at least 150 mm above the 100-year floodwater level and/or the worst recorded historic flood level.

3.8.2 River characteristics remain unaltered

A frequent requirement of T&CPA approval is that the development should not affect the river characteristics and that net flood storage area is not reduced.

3.9 Industrial and commercial sites

3.9.1 Separate sewer systems and light liquid separators for runoff from polluted hardstandings

Some industries require heavy goods vehicle forecourts where pollutants are deposited then washed off during rainfall. For such locations, it is important to develop individual drainage systems: one for roofs and other areas not subject to potential pollution, and the other served by light liquid separators or passed to treatment before discharge.

3.9.2 The use of effluent treatment

Some sewerage undertakers will require pre-treatment or charge higher connection fees for receiving effluent that has a significant impact on the wastewater treatment works.

3.9.3 Grease separators

Grease causes severe problems at sewage treatment plants, particularly smaller ones, in pumping stations and pipework generally. Grease separators or other control measures should be installed on discharges from all commercial catering or food processing facilities.

Similarly, macerating solid wastes and discharging the output to sewers causes unnecessary problems and costs and is considered environmental bad practice. Solids or semi-solid waste should be disposed as a solid waste; compactors are available to facilitate handling.

3.10 Landscaping

Landscaping includes two issues that relate to drainage design.

3.10.1 Runoff control and location

The location of storage, temporary flooding and controlling flood flows to prevent flood damage is dependent on existing topography, planned landscaping and the development's layout. Specific attention to detail considering what might happen if an extreme event hits the site is an important exercise in the initial design of the site.

3.10.2 Runoff volume

The addition of large areas of impermeable surfaces results in increases in both the volume and the flow rate of surface water. This is particularly true for sites on sandy soils (see Section 3.6). However, preventing runoff by using areas of the catchment for infiltration or using landscaping for temporary ponding can minimise the effect of the volumetric impact and so reduce storage needs as part of drainage discharge requirements.

3.11 Environmental issues

T&CPA authorities have begun imposing sustainable drainage (SUDS) practices. This is aimed at reducing the detrimental impact of urbanisation on rivers and groundwater. It is therefore important for developers to be aware of the environmental issues that apply on each site (see Figure 3.1).

3.12 Contaminated land

Contaminated land (CIRIA SP124 – Privett *et al*, 1996) brings up numerous issues with which the developer must deal, although few specifically relate to drainage. The principal concern is the impact of pollutants on groundwater. The use of soakaways and perhaps other infiltration-related drainage methods would not be acceptable in these situations due to the migration of pollutants in the ground.

3.13 Size of development

The size of a catchment does not affect the principles of drainage design, layout and analysis. Where storage is being considered, however, there is a practical limitation in applying throttle limits as sewerage undertakers rarely accept pipe sizes that are smaller than 150 mm. This results in a limit of discharge that generally must be greater than 10 l/s.

3.14 Water reuse

The pressure on water resources makes the reuse of rainwater and wastewater an attractive option – one that is being considered in many parts of the world including the UK (see CIRIA PR80 – Leggett *et al*, 2001). Although there are great benefits to be gained, using wastewater for domestic and industrial purposes also presents risks and difficulties. Wastewater covers several categories, some of which are more viable than others. They are usually classified as:

- rainwater
- greywater
- black water.

3.14.1 Rainwater

Rainwater is probably the most attractive of the alternative water supply options. Many households already use it for their gardens, but it is now also being considered for replacing or augmenting treated water in toilets. More adventurous proposals envisage rainwater augmenting treated water for all household uses. Generally, the water is collected from roofs in a small tank on the ground and pumped up to recharge the cistern as required.

Theoretically, this not only lessens consumption of precious treated water, but also reduces runoff impact on sewer systems, thereby minimising combined sewer overflow discharges. As roofs represent around 50 per cent of drained impermeable surface areas in combined-sewer catchments, the benefits can be significant.

The drawbacks to this approach are detailed below.

Sustainability
Current house design generally prevents the use of gravitational systems for rainwater collection and so requires the less sustainable process of pumping it up to header tanks. A completely gravity-based system avoids this problem, but requires changes in house design to allow water to be stored in lofts.

Treatment
The health risks of using rainwater mean that it has to be treated. As gutters collect leaf litter and bird droppings, both the solids and bacterial content must be considered and designed for. If not treated, the water is likely to cause odours and present a significant risk to health, particularly that of pets. The legal responsibilities would be a matter of concern for water supply companies and the onus would be on the property owner to assume responsibility for this aspect.

Local collection, treatment and distribution
Local collection, treatment and reuse of water are occasionally proposed. The advantage of scale is a major bonus, as is the additional protection of having a central treatment facility. However, the secondary collection and distribution system results in a plethora of pipework and rarely makes it a cost-effective approach.

Design of sewers
Sewer design downstream, although theoretically benefiting from reduced runoff, could not rely on reduced flows, as the storage systems might be full. The relationship between tank size, available storage and design storm events would have to be very carefully assessed. It is likely that a conservative set of assumptions would be made to avoid the risk of under-designing the collecting system.

While technically all problems can be overcome, this approach is likely to be limited to areas where water is scarce.

3.14.2 Greywater

Greywater is defined as wastewater from baths and hand-basins and sometimes also washing machines and kitchen sinks. Like rainwater, greywater presents bacterial and odour problems. There are two main advantages of a greywater system over a rainfall system: chiefly, the continuous availability of the resource, which allows tank sizes to be smaller; secondly, the solids element is reduced as rainwater runoff can include a variety of detritus collected in gutters. However, the bacterial and odour issue is generally more of a problem in greywater than rainwater reuse.

3.14.3 Black water

Black water generally refers to toilet wastewater or all domestic wastewater. One or two locations in the world have collected and treated sewage and used it to augment water supply by mixing after tertiary treatment. This is done only by water authorities at the catchment level and solely because of the pressure on water resources in those areas. Local or domestic reuse of black water is never considered.

3.15 Land drainage

Land drainage, whether by means of ditches or subsurface land drains, is generally not picked up by drainage systems when developing sites. Both the Environment Agency and the sewerage undertakers usually require it to continue to be discharged to the stream or watercourse to which it passed before development.

3.16 Foul/roof/road pipe systems

3.16.1 Adopting authorities

New developments are nearly always built with a twin-pipe system, one for foul water and the other for surface water. Often these are both connected on the boundary to a single manhole, as the main collecting system may be a combined sewer. Sometimes there is pressure to have a three-pipe system, as the sewerage undertaker is liable only to serve properties and the Highways Agency is responsible only for providing road drainage. In practice, it is generally agreed that a three-pipe system is impractical, so where roofs are drained by a pipe system, the sewerage undertaker usually agrees to have the road drainage included. There are exceptions and special cases due to specific catchment characteristics and site layouts, but it is important to agree the adoption of the pipe systems with the relevant authorities. If there is a lack of agreement between various parties, this can be resolved by appealing to the Office of the Deputy Prime Minister (ODPM).

3.17 Vacuum sewerage

Vacuum sewerage is rarely used, but is suited to places where local constraints make traditional gravity drainage difficult to implement or particularly expensive. It is a specialist area of drainage design. Appendix B provides information on the design and use of such systems.

3.18 Strategic installations

Attention should be paid to site security and the potential for vandalism with regard to all drainage infrastructure. This would be particularly necessary at, for example, power stations, the Channel Tunnel and the London Underground.

3.19 Areas of risk particular to site drainage

Site drainage construction risks relate particularly to confined spaces and trenching. This guide is not aimed at detailing health and safety issues. CIRIA C604 (Ove Arup and Gilbertson, 2004) defines risks in various categories. It is recommended that engineers obtain this book, or a similar document, to assist them in implementing the CDM requirements. It not only highlights possible hazards for each category of activity, but it also provides HSE references that are relevant to each activity.

3.20 Detailed design issues

In addition to these general considerations, specific site characteristics also need to be taken into account. Figure 3.1 summarises some issues that need to be considered during design of site drainage that will affect decisions related to drainage. Figure 3.1 should only be treated as a guide. All design issues should be considered when designing drainage systems for a site.

DESIGN ISSUES	INFLUENCING FACTORS														
	Development category				Environmental factors					Site characteristics					
	Industrial	Commercial	High density residential	Low density residential	Land contamination	Groundwater sensitivity	River water quality	River low flow regime	River high flow regime	Soil type	Size	Hydrology	Topography	Existing sewerage	Existing drainage
Runoff volume	✓	✓	✓	✓				✓	✓	✓	✓	✓			✓
Runoff rate	✓	✓	✓	✓				✓		✓	✓	✓	✓		✓
Effluent characteristics (pre-treatment)	✓	✓				✓	✓							✓	
Grease separators	✓	✓												✓	✓
Light liquid separators	✓	✓				✓	✓								
Soil treatment					✓										
Infiltration	✓	✓	✓	✓	✓	✓	✓		✓	✓		✓			
Interception	✓	✓				✓	✓				✓	✓		✓	
Limited discharge settings	✓	✓	✓	✓				✓	✓	✓	✓	✓			✓
Storage capacity	✓	✓	✓	✓			✓	✓	✓	✓	✓	✓			✓
Location (roads, stormwater storage, buildings)													✓	✓	✓
Connection levels													✓	✓	✓
Downstream impact							✓	✓	✓						
Overland flow										✓	✓	✓	✓		
Storage design	✓	✓	✓	✓			✓	✓	✓	✓	✓	✓			✓
Network design	✓	✓	✓	✓						✓	✓	✓	✓	✓	✓
Types of drainage	✓	✓	✓	✓	✓	✓	✓	✓	✓	✓		✓	✓		

Figure 3.1 *Issues influencing site drainage design*

4 GENERAL INTRODUCTION TO DRAINAGE PRINCIPLES

4.1 Types of drainage system

Normally, new site developments in the UK have to be provided with separate drainage systems for the foul and surface water flows, namely:

- **foul water systems**, which deal with sanitary and wastewater discharges from buildings and with other flows from a site that cannot be discharged to watercourses or groundwater without prior treatment
- **surface water systems**, which carry surface runoff (usually but not exclusively produced by rainfall) from roofs of buildings, paved surfaces, car parks and, sometimes, contributing natural areas.

Under Section 104 of the Water Industry Act 1991, the sewerage undertaker may agree (on payment of appropriate fees by the developer) to "adopt" some or all of the foul or surface water sewers within a new scheme as part of the public sewerage system (see Chapter 2). In most cases, only those lengths in public roads or in areas where the sewerage undertaker has rights of access will be adopted; other parts of the drainage system will normally remain the responsibility of the owner of the site or of the purchasers of individual properties. Pumping stations within a site may be adopted provided the access rights and ownership of the land that they occupy are transferred to the sewerage undertaker. Criteria covering the design and construction of sewers that are intended to be adopted as public sewers have been produced by the Water Services Association and published as *Sewers for adoption* 5th edition (WRc, 2001).

In the case of an addition or modification to an existing building development, it is sometimes be acceptable to use a partially separate drainage system in which runoff from roofs (and, possibly, also small paved areas) is connected to the foul water system. However, this can only be done with the agreement of the sewerage undertaker if a connection is to be made to the public sewerage system.

In most situations it will be necessary to provide separate drainage systems within a site even if they discharge into a public combined sewer that carries both foul and surface water flows. The separate systems may be connected at the last manhole within the site, but the surface water drain should be trapped to prevent smells and possible migration of explosive gases from entering it from the foul system.

Drainage systems that will not be adopted as public sewers are covered by Building Regulations and are subject to approval by building control. Designs that satisfy the criteria in relevant British and European Standards or other approved documents are deemed to satisfy the Building Regulations (see Chapter 2).

4.2 Overall design criteria

Site drainage systems should be designed to provide a specified degree of security against surface flooding or against surcharging of a below-ground piped system. The stages to be followed in the hydraulic design of any type of drainage system can be generalised as follows.

1. Specification of the required level of performance for the system based on the degree of security needed against surcharging or surface flooding.
2. Calculation at key points in the system of design rates and volumes of flow whose probability of being exceeded by rare events is acceptably small and consistent with the required level of performance.
3. Sizing of the components of the drainage system at these key points to provide the required values of flow capacity or storage.
4. Checking of the hydraulic design to ensure that other criteria affecting performance are satisfied (eg minimum pipe sizes or minimum values of flow velocity needed to produce self-cleansing conditions).

Detailed information and recommendations for the different drainage systems and components are given later in this guide. However, general issues concerned with levels of performance and degrees of risk are dealt with in Section 4.3.

4.3 Risk criteria

If an event (eg the occurrence of a certain amount of rainfall in a particular time) occurs or is exceeded on average once every T years, it is said to have a return period of T years. The probability, P_r, of a more severe event occurring in the design life, L (in years), of the building or system is given by the formula:

$$P_r = 1 - [1 - (1/T)]^L \qquad (4.1)$$

The value of P_r can vary between 0 and 1, the latter indicating that there would be a 100 per cent chance of the event being exceeded during the design life. As an example of the use of Equation 4.1, it can be shown that there is a 40 per cent chance that a design storm with a return period of T = 50 years will be exceeded in severity once during a period of L = 25 years (ie, the probability is $P_r = 0.40$).

Developers, drainage designers and owners of buildings need to be aware that it is practically impossible to guard against all possible risks of flooding from drainage systems. As shown by Equation 4.1, there will always be some finite risk that the design flow conditions will be exceeded and that some inconvenience or damage will result. As in the case of other hazards such as wind, earthquake and fire, it is necessary to find an economic balance between:

- the cost of constructing the system
- the costs incurred if the capacity of the system is exceeded
- the level of probability of this occurring during the design life of the building or site being drained.

4.4 Governing risk in component design

The balance between the costs and probability of damage described in Section 4.3 will vary from one part of a drainage system to another. Thus, the consequences of a roof drainage system overflowing and causing flooding inside a building are likely to be far more severe than any inconvenience resulting from temporary flooding of an outdoor car park. As a result, it would normally not be appropriate or economic to apply the same risk criterion to all parts of a site drainage system. The principle to be applied in such cases is that of governing risk, as explained below.

If a lower level of hydraulic performance is appropriate in one part of a system than another, this is acceptable provided there is no adverse effect on the one that is more critical. If this is not the case, the risk criterion applying to the more critical part should be the governing factor for both components of the system. The tendency to allocate design responsibilities for different parts of a drainage system to specialist engineers can lead to these issues being overlooked, so it is important that the co-ordinating designer for the scheme takes an overall view of the whole drainage system and checks that there are no inconsistencies in the design assumptions for the various components.

An example of the application of governing risk is the interface between a rainwater drainage system for a building and the below-ground piped system collecting surface water from the site. If a conventional roof drainage system has vertical rainwater pipes outside the building which flow only partly full, there is no mechanism by which water levels in the below-ground pipes can affect flow conditions at roof level. Therefore, there need not be any inconsistency if the gutters were to be designed for a storm return period of 100 years while the below-ground pipes were sized for storms with a return period of one year. However, if the vertical rainwater pipes were located inside the building and connected to a drainage system beneath the floor, there would be a danger of internal flooding of the building. This could occur because, in rarer storms, the roof drainage system would be capable of dealing with much higher rates of flow than the external below-ground pipes; the latter might become temporarily surcharged and cause water to back-up into the building. In this situation, the principle of governing risk would require the piped drainage system close to the building to be designed for a higher degree of security than would normally be the case. Similar types of problem can occur if the roofs of buildings are drained by siphonic systems (see Chapter 5), or if the risk criteria for storage tanks or infiltration systems are incompatible with those for the below-ground pipes that serve them.

General recommendations on appropriate risk criteria for different parts of drainage systems are indicated in Figure 4.1 and discussed in Chapters 5–10, but it should be appreciated that there is no single set of rules that will be valid for all situations. Each scheme needs to be considered separately, taking account of the particular design requirements and layout of the site.

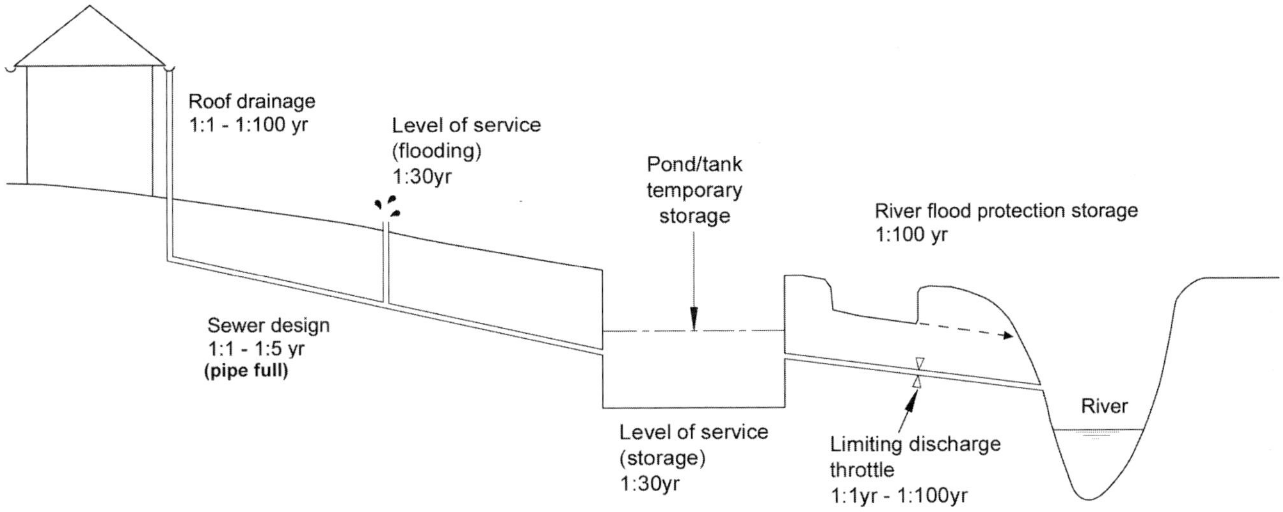

Figure 4.1 *Schematic of risk applied to component design – roof and site drainage*

4.5 Design rates of flow

4.5.1 Foul water systems

At the upstream end of a foul water system the flows tend to be peaky and intermittent because they are produced by the operation of individual appliances within buildings. Farther downstream, as the number of connections to the system increases, the flows become more continuous but still subject to daily peaks corresponding to patterns of domestic or industrial usage.

The design flow rate at a point in a foul water system depends on:

- the number of water appliances (WCs, sinks, baths, showers, industrial equipment etc) within the buildings served
- their flow characteristics (volume and duration of flow)
- their frequency of use.

Based on this information, it is possible to calculate the probabilities of different numbers of appliances being operated simultaneously. Design recommendations described in Chapter 5 are based on limiting the chance that the design flow rates will be exceeded to 0.5 per cent of the time (ie, a probability of $P_r = 0.005$; see Section 5.3).

The design flow rates of foul water systems that are to be adopted as public sewers are calculated for domestic housing by assuming an average daily rate of flow per dwelling (based on assumed numbers of occupants) plus, if appropriate, an allowance for infiltration into the system. This flow rate is termed the **dry-weather flow** (DWF). The peak daily flow rate in a foul water sewer is typically equal to around 2 × DWF (higher at upstream parts of the system and less downstream as translation and attenuation takes place). New developments generally form the upper branches of the sewerage network, therefore recommended practice from *Sewers for adoption* (WRc, 2001) is to design the pipes so that they have a flow capacity of 6 × DWF when flowing full. This allows for future developments, misconnections and infiltration. Full details are given in Chapter 9.

For large industrial or commercial buildings, a more detailed analysis is necessary to establish the values of DWF and the peak flow characteristics.

4.5.2 Surface water systems

Flows in these systems are far more variable in terms of magnitude, duration and frequency than flows in foul water systems. The processes involved in the runoff of rainfall from buildings, paved surfaces and unpaved areas within a site are very complex, and the degree of detail that needs to be considered in design depends on the scale of the problem and on the likely severity of the effects of flooding.

Nearly all methods for sizing surface water drainage use the following type of equation to calculate the design rate of flow at a point in the system:

$$Q = f\,(F_{RP}\ C_V\ C_R\ A\ I) \qquad (4.2)$$

where:

- Q is the flow rate
- F_{RP} is a function related to return period
- C_V is the runoff coefficient for the catchment (non-dimensional; a value of $C_V = 1.0$ means that all the rainfall falling on the catchment produces runoff at the point in question)
- C_R is a routeing coefficient (non-dimensional) that takes account of storage effects in the drainage system and shape of the catchment
- A is the effective plan area of the catchment (often taken as being the impermeable area)
- I is the rainfall intensity.

Flow is therefore a function of intensity and return period. Design flow rates used to size drainage systems are calculated based on a flow rate using either a method called the Rational Method or the Hydrograph Method. The peak rainfall intensity increases with higher return periods and decreases with the increase in the duration of the design storm. The design storm for analysis increases as the storm duration reduces.

It should be clearly understood that designing for roof drainage for a 1 year 2 minute event may predict a required flow rate that is higher than a 5 year 1 hour event which might be used to design the surface water drainage system.

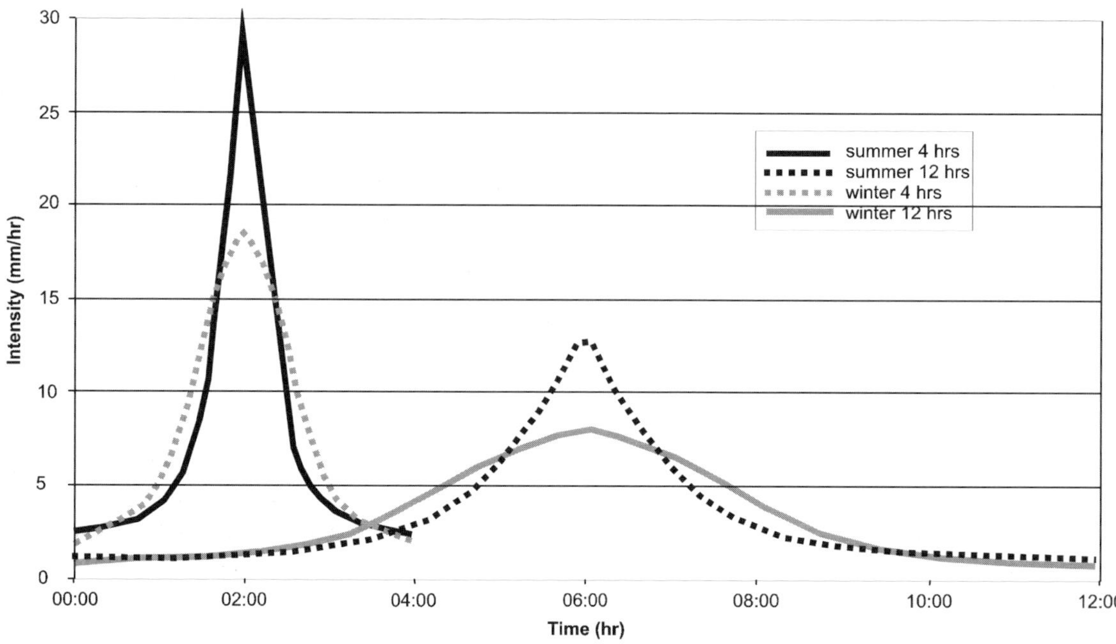

Figure 4.2 *Five-year summer and winter rainfall hyetographs – four-hour and 12-hour durations*

There is a hierarchy of methods used by engineers to design drainage systems. These can be summarised as below.

Outline design

- A constant value of rainfall intensity is used.

This is based on a rule-of-thumb approach founded on experience that allows approximate pipe sizing and levels to be established. A constant rainfall intensity is used to determine flow rates at all points in the system. This approach is used particularly for small sites and a check is subsequently made at the detailed design stage to confirm that sizes are adequate. Different constant intensities are used depending on either the system being designed or the location in the country/world being considered. Thus a constant intensity used for designing the gutters would differ from that used for the surface water drainage system, or one part of the system needed to be designed to a higher specification than another.

Drainage of small areas (up to 2000 m^2) is usually calculated using a flat rate of rainfall, generally related to the worst five minutes of a storm event; the formula taking the form:

$$Q = AI \tag{4.3}$$

Where rainfall intensity is in $l/s/m^2$, flow is in l/s if area is applied in m^2.

The Rational Method

- Rainfall intensity varies with return period and duration. The concept of "time of concentration", T_C is used to determine the rainfall intensity for any duration.

The Rational Method has been used for many years (Mulvaney, 1850 and Lloyd-Davis, 1906) and allows a relatively accurate assessment of flow, and therefore pipe size requirements, for any system that is not too large (covering less than 100 ha) as long as rainfall data exists to derive rainfall-intensity-duration curves.

One of the approaches used for sizing surface drainage systems is the Rational Method. This method assumes that, for a given frequency of occurrence, the greatest possible flow rate will occur when the duration, D, of the storm is equal to the value of T_C for any point on the network. In the simpler versions of the method, the coefficients C_V and C_R in Equation 4.2 are defined by a common coefficient C and is often taken to be a value of 1.0 for all impervious surfaces. In the Modified Rational Method (HR Wallingford and IH, 1981b) the routeing coefficient has a value of $C_R = 1.3$ and the runoff coefficient, C_V, has a value that is typically between 0.6 and 0.9 depending on the quality of the impermeable surface in the catchment.

The **time of concentration**, T_C, at a point in a drainage system is the time taken for all the catchment upstream of that point to contribute runoff. For a below-ground piped system, TC is equal to the time taken for overland flow to enter at the head of the system (the **time of entry**) plus its time of travel along the pipes.

Hydrograph Method

- The rainfall intensity varies with time during the design storm. Design storm events have different intensity profiles depending on storm duration and return period.

The main difference between the Hydrograph Method and the Rational Method is that the analysis is carried out using the volumes of runoff and the volume available in the pipe system, which first was taken into account in Road Note 35 (TRRL, 1976). For large systems the volumetric and routeing effects makes the Hydrograph Method more accurate than the Rational Method. In addition, the Hydrograph Method allows surcharge analysis – and therefore testing for flooding and its potential impact (see Section 9.5) – to be carried out.

4.6 Hydraulic design of pipe systems

When designing a pipe drainage system to cater for the required flow rates, it is necessary to have the following information:

- a resistance equation for pipe flow that relates the flow capacity to the internal size of the pipe, the proportional depth of flow in the pipe, the gradient at which the pipe is laid, and the resistance characteristics of the pipe walls. The same type of equation can also be used to calculate frictional losses in pumping mains
- allowable minimum (and possibly maximum) pipe sizes
- minimum and maximum pipe depths
- minimum or **self-cleansing velocities** in the pipes that need to be achieved at regular intervals in order to prevent the build-up of sediment deposits and blockages.

The resistance equation recommended in all the design guides, which is described in Chapter 9, is the Colebrook-White equation. Full details of this are given in Appendix F. The resistance characteristics of the pipe walls are described in terms of an equivalent surface texture height, ks, which is usually expressed in millimetres. Clean, new drainage pipes can have values as low as ks = 0.03 mm. The walls of pipes carrying foul sewage rapidly become coated by biological slimes and greases, however, and typical in-service values for gravity drains can be between about 0.6 mm and 1.5 mm. This increases to around 3.0 mm when taking into account minor cracking and jointing offsets for older systems that are considered to be in reasonable condition. Sewage pumping mains tend to have somewhat lower values of ks because the sliming is thinner and more uniform around the walls. Surface water sewers do not normally experience sliming, but sediment deposits on the invert can produce large increases in the effective resistance of a pipe. Part of the increase is due to the loss of flow area and part due to the surface roughness of the sediment bed; values of ks can reach 30 mm or more. Application of suitable minimum values of flow velocity when designing drainage systems can help minimise the adverse effects of deposits.

Recommendations on these various factors are given in Chapter 9 of this guide.

5 ROOF DRAINAGE

There are two main types of roof drainage system available in the UK and in other countries: conventional and siphonic. They require substantially different design approaches.

Inadequate performance of roof drainage systems is often at the root of internal flooding incidents affecting commercial, industrial and retail buildings with large covered areas. Some of the causes of rainwater flooding in buildings can be considered as "natural hazards", such as when the rainfall intensity in a storm exceeds the level considered to be economically viable to guard against, but other causes can be eliminated by careful consideration of the type of roof drainage system, appropriate design and construction. Regular maintenance to remove leaves, dust and other debris that tend to accumulate at roof level is also essential to ensure unrestricted rainwater discharge. This chapter details:

- types of roof drainage system
- standards
- criteria for design
- limitations of current regulations.

5.1 TYPES OF ROOF DRAINAGE SYSTEM

The two main types of roof drainage system (see Figure 5.1) are described overleaf.

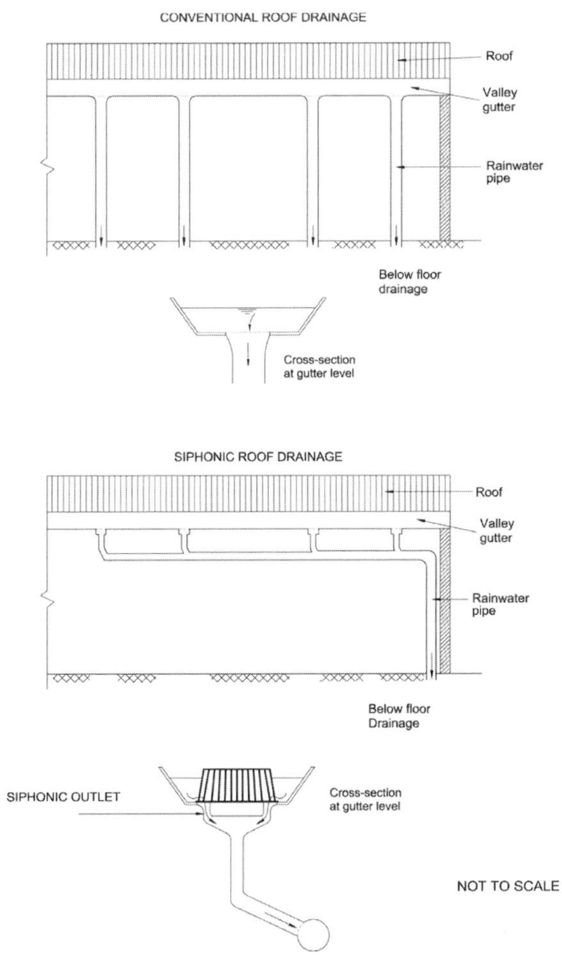

Figure 5.1 *Conventional and siphonic roof drainage systems (schematic diagrams)*

Conventional

These systems are formed by single outlets (usually circular openings) in the roof or sole of a gutter that connect to vertical rainwater pipes (or downpipes). They are designed to flow part-full under atmospheric pressure following the principle that the water depth around the outlet builds up and weirs through the outlet into the rainwater pipe. The head available to drive the flow is the depth of water around the outlet. The larger the flow rate, the deeper must be the head over the outlet. Increasing rainfall can cause the outlet to drown and from this point onwards a much greater increase in water depth is required to discharge the additional flow. Current regulation recommends that the pipework for conventional systems should be designed to run at most only one-third full to avoid the problem of excessive depth of water on the roof or gutter, with the associated risks of unacceptable load on the roof and of internal flooding of the building.

Advantages
Greater simplicity of design; lower vulnerability to limited maintenance.

Disadvantages
Require one downpipe per outlet, which can be a severe limitation in large buildings designed to provide large uninterrupted internal spans.

Applications
Conventional systems are best suited for buildings with a small roof, such as in housing developments and small retail and industrial outlets. They can also be suitable for larger buildings that have few constraints regarding the location of the downpipes and the associated underground pipework inside the building, as well as several points of discharge into the site drainage system.

Siphonic

These systems are able to discharge significantly higher flow rates than conventional systems because they are designed to utilise the full difference in height between roof level and the point of discharge at or near ground level. Although they can consist of a single, specially designed outlet (siphonic outlet) connected to a downpipe, to take full advantage of their high capacity, most systems consist of several outlets connected to a single downpipe. The principle of operation relies on limiting the amount of air entering the system and on the expulsion of air (or a large proportion of the air volume) from the pipework (priming). Siphonic outlets incorporate a baffle or plate to induce priming and the siphonic action in the pipework. This demands a completely different design of pipework from that of conventional systems, involving quite complex computational analysis. The specification of the pipework has to take into account that sub-atmospheric pressures that can occur in the pipes. Since siphonic systems are able to drain very large areas through a single downpipe, there is potential for large flows being discharged into the site drainage system at each drainage point.

Advantages
Ability to drain very large areas with few downpipes, thus reducing the need for vertical pipes and below-ground pipework inside the building.

Disadvantages
Design of siphonic systems requires the use of specialised software and is usually carried out by manufacturers/suppliers. Due to its technical complexity and the large flow rates involved in many cases, it is recommended that the design be independently checked. Since the pipework is smaller in diameter than conventional ones, siphonic systems tend to be more dependent on frequent maintenance to ensure adequate performance.

Applications
Siphonic systems often provide the most effective means of draining large roofs of industrial, commercial and retail buildings, airport and railway terminals. They are particularly useful when, for architectural or other reasons (such as the availability of site drainage discharge points), the number and size of down-pipes need to be kept to a minimum.

5.2 Standards

BS EN 12056-3:2000 *Gravity drainage systems inside buildings. Roof drainage, layout and calculation* is the European standard dealing with the design of roof drainage. It supersedes British Standard BS 6367:1983 *Drainage of roofs and paved areas* and incorporates six national annexes containing specific information retained from BS 6367, which would otherwise become unavailable. These annexes have an "informative" status and take account of traditional UK practice. Among the topics covered are new rainfall maps of the British Isles, procedures for the design of gutters with restricted discharge and testing of siphonic outlets.

The Standard covers the design of conventional roof drainage systems and gives some general performance guidance on siphonic systems. Also included in the Standard are normative guidelines for the testing of gutters and outlets. The current Standard has a different calculation method for conventional roof drainage systems when compared with the superseded BS 6367, but they both have the same theoretical and experimental basis. It is also worth noting the change in the units and symbols used: for example the rainfall intensity is given in l/s per m^2 (as opposed to mm/h).

The hydraulic design of roof drainage systems, even if only of the conventional type, can be fairly complex and the formulae and procedures recommended in the Standard are more suited for calculation using proprietary software specifically designed for this purpose.

5.3 Criteria for design

1. Criteria that apply to both conventional and siphonic systems:

- rainfall intensity, which depends on:
  - **duration** of the rainfall event (taken as equal to the time of concentration, ie the time for the rain falling on the most upstream part of the roof to reach the outlet from the roof or roof gutter)
  - **geographical** location of building (maps are given in the Standard showing contours of rainfall intensity for events with two-minute duration and return periods of 1, 5, 50 and 500 years up to the maximum probable rainfall). Estimation of storms of other durations up to 10 minutes is also covered in the Standard
  - **return period** of the event based on the category of the building. There are four categories in BS EN 12056-3: Category 1 for eaves gutters and flat roofs and return period of 1 year, and Categories 2 to 4, which are defined in terms of the probability of the design rainfall intensity being exceeded during the life of the building.

Table 5.1 *Rainfall criteria for the design of roof drainage*

Category	Probability of rainfall being exceeded over life of building	Design return period (years)	Comment
1	–	1	Eaves gutter
2	0.5	1.5 × design life of building	
3	0.2	4.5 × design life of building	
4	$\rightarrow 0$	Maximum probable rainfall	

- effective catchment area – takes the effect of wind into account.
- design flow loads – this is the flow that will enter the site drainage system.
- gutter design/outlet capacity – if the outlet prevents the gutter from discharging freely, the two components need to be considered in combination.

2. Criterion that applies only to conventional systems

   - The limiting capacity of vertical rainwater pipes is based on the assumption that they will run part-full at 33 per cent of its capacity.

3. Criteria that apply only to siphonic systems

   - As mentioned above, the design of siphonic systems follows specific hydraulic principles that must be adhered to. However, certain aspects, such as the choice of pipe diameters in different parts of the system, also rely on the experience of the designer. Other aspects, such as the time required for the system to prime, ie to start working siphonically, at the beginning of a storm event, are still being investigated. Some established criteria specific to siphonic systems are:
     - the capacity of the system depends primarily on the vertical height between the roof and the point of discharge
     - the storage capacity of gutters should not be taken into account in determining the capacity of the system
     - systems formed by several outlets should be designed to balance the flows through each outlet
     - minimum velocities should be achieved in the pipework for the design flow rate to minimise the risk of deposition in the pipes and to speed up the priming process
     - the design calculations should include the determination of pressure values inside the pipework. Negative pressures should be sufficiently below the vapour pressure of water everywhere in the system to avoid the risk of pipe implosion and cavitation (the damage caused by the collapse of air cavities near the pipe walls)
     - the design needs to take into account the conditions at the point of discharge into the local site drainage (free or surcharged).

5.4 Limitations of current regulation and their implication for site drainage

The current Standard, BS EN 12056-3:2000, provides guidance on design of conventional roof drainage and on estimation of flows for both conventional and siphonic systems. These latter systems, because of their larger capacities and flow velocities at the discharge point, can impose a severe burden on the site sewer network. Care is required in designing the discharge of the high-velocity flows from siphonic systems safely into the sewer. The provision of chambers to dissipate part of the energy of the flow may be one of the features that need to be considered.

According to the Standard, the roof drainage of a building can be designed for return periods that are much longer than those generally considered for the local site drainage sewer (in some cases more than 100 years). There is need to consider this difference in return periods in designing the roof drainage system.

The scarcity of design guidance on siphonic systems and their interaction with the site drainage is perpetuated in the Standard BS EN 12056-3. This is essentially due to the complexity of these systems and to the difficulty in reconciling the different design approaches adopted by the system manufacturers.

Roofs of unusual shape often also require solutions for the removal of rainwater (for example, high-level energy dissipators or chutes that are unlikely to be covered by the Standard. The hydraulic design of these systems will therefore need to be dealt with by specialist hydraulic consultants. With regard to siphonic systems, effects such as high-velocity jets entering the site drainage should be taken into account.

6 BASEMENT DRAINAGE

Basements require particular attention as, being below ground, there is perhaps greater potential for them to suffer from flooding as well as creating problems in serving them using gravity drainage systems. In practice it is not only basements that need to be considered as hydraulically surcharged sewers can cause flooding inside a property even though floor levels are above ground. Basements are rarely constructed in the UK in a domestic context, though many countries in Europe use basements for car parking as a matter of course. Multi-storey car parks and some industrial sites do build below ground, however.

This chapter draws attention to:

- analysis for basement flood protection
- methods of protection.

6.1 Analysis for basement flood protection

As with other internal drainage from buildings, basements are drained to the foul system. All modern developments are built using separate systems. Although foul systems are, in theory at least, not affected by rainfall, it is normal practice to connect a basement to a foul system pipework using a pump. The use of a gravity system would normally be inappropriate because, even if the foul system was built below the basement floor level, there is a tendency over time for foul systems to respond to rainfall, due to misconnections or general infiltration, and for the system to operate in surcharge with the consequent risk of internal flooding.

Basements can be affected by seepage problems if the tanking is not totally watertight. Because of this, it is quite normal to allow for drainage even if there are no ablution facilities – see CIRIA Report 140 (Johnson *et al*, 1995).

Many developments are infill areas within cities that have combined systems. In these circumstances, the surcharging effects that can be expected in extreme events must be specifically considered. It is therefore important to look at the hydraulic gradients for networks, particularly for areas with basements, to evaluate whether any property is at risk. Figure 6.1 illustrates how the hydraulic gradient can cause basement flooding. It is important to establish whether the basement is connected to the network and whether this is a gravity link.

The foul system would usually be assessed on the basis some degree of misconnection (additional area contributing rainfall runoff) based on experience. It is generally recognised that 3–5 per cent of the impermeable area in a catchment will be misconnected in a mature estate.

6.2 Methods of basement protection

Generally basement protection is a problem that has to be faced for mature sites, usually in areas served by combined systems. Flooding of basements should not generally be a problem for new sites although, as previously stated, infill areas need to be specifically considered for potential hydraulic problems.

There are two main options to protect the property.

1 **Anti-flooding device** (AFD) – to prevent backflow from a public sewer.

  An AFD may have an automatic closure device (eg a hinged flap that closes when the flow direction reverses), or a powered closure device (eg an electrically operated flap that is linked to a water-level sensor and an alarm), or a combination of the two.

2 **Pumping** – to pump drainage flows from a property into a public sewer.

  The pump will normally be installed in the basement and discharge to the public sewer either under pressure or via a gravity pipe.

Although the industry uses AFDs, they are usually categorised as temporary solutions, so it is unlikely that a new development would utilise these units except in special circumstances. Where their use is proposed, and there is solids material in the waste, it is recommended that powered closure versions be used.

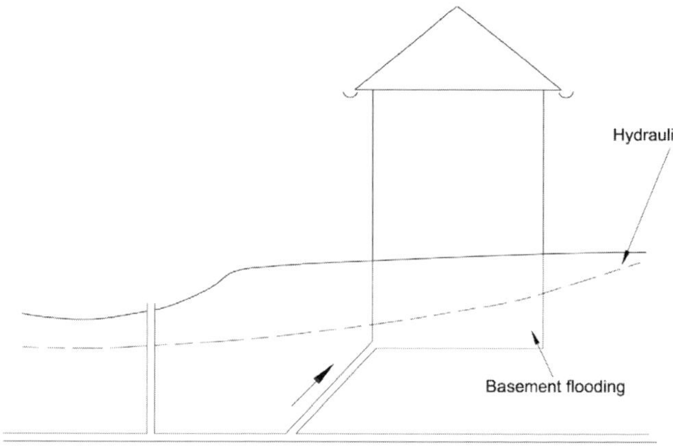

Figure 6.1 *Potential for property damage due to surcharged flows in drainage systems*

Pumps are more commonly designed into new structures where there is a need to cater for these situations. In certain countries it is mandatory to provide a pump in a basement.

Basements are regularly built in many countries in Europe and the proposed BSI Normative Standard BS EN 12056-4: 2000 *Gravity drainage inside buildings. Wastewater lifting plants – layout and calculation details methods of designing and calculating against backflow of sewage into buildings*. Related Standards include BS EN 12050-1 to 4: 2001 *Wastewater lifting plants for buildings and sites. Principles of construction and testing*, which addresses the use of pumps and non-return valves.

For both of these options (pumps and non-return valves) there is an inherent maintenance requirement needed to ensure that they remain in good condition.

The following Figures 6.2, 6.3 and 6.4 are excerpts from BS EN 12056 and illustrate the options considered for basement drainage.

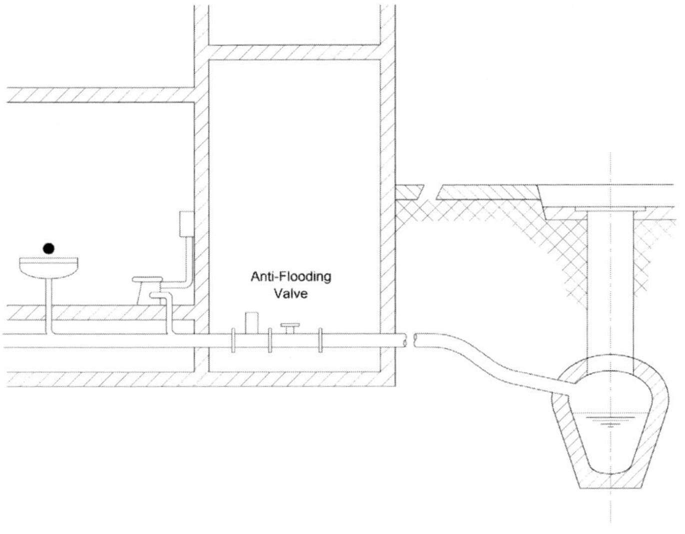

Figure 6.2 *Protection against backflow by means of anti-flooding valve where there is a natural fall to the sewer*

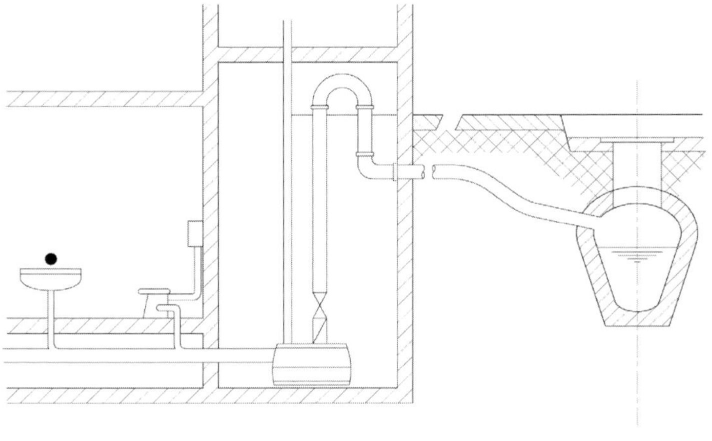

Figure 6.3 *Protection against backflow by means of a pump where the sewer is higher than the sanitary appliance*

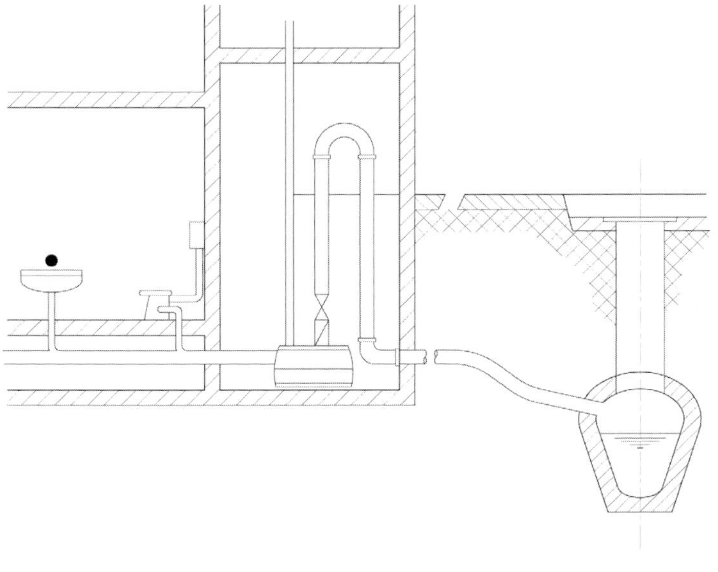

Figure 6.4 *Protection against backflow by means of a pump where there is a fall to the sewer*

7 PAVEMENT DRAINAGE

This chapter is concerned with the drainage of paved areas that are used as car parks, pedestrian precincts and roads serving the development site. Pavements within the property curtilage are often constructed in a similar manner to pedestrian precincts, but are rarely specifically drained unless falls are away from the road. Pavements can be constructed so that they either provide an impermeable or a permeable surface. Their more limited bearing capacity means that permeable (porous) pavements are usually restricted to low-density traffic areas and are a form of infiltration system – see Chapter 8 for more detail.

Permeable pavements should be distinguished by subdividing into two types. The first is where a permeable wearing course is applied to a standard road for noise and spray reduction; the second being a porous construction allowing water to pass through into the sub-base.

Drainage can be carried out by conventional methods, such as kerb and gully systems or drainage channels, but sustainable drainage methods (SUDS) are increasingly being specified and built in an attempt to reproduce the natural runoff and infiltration that takes place in the non-built environment. Reference to sustainable systems are presented in this chapter, but more detailed information is provided in Chapter 8.

This chapter covers the following items:

- options for pavement drainage
- standards
- layout
- types of pavement drainage
- design principles.

Pavements are defined here as surfaces that are usually built at ground level for a range of purposes such as enabling vehicular access to buildings (site roads), provision of firm surfaces for parking of vehicles (car parks) and access and amenity areas for pedestrians (pedestrian precincts). A distinction is made in the following sections between site road drainage and the drainage of car parks and pedestrian pavements.

7.1 Options for pavement drainage

7.1.1 Site road drainage

Drainage of roads serving a site development can be achieved either by conventional means or by sustainable drainage (known as SUDS). An overall listing of these systems is given in this section but they will be described in more detail in Chapter 8.

Conventional road drainage is not limited to kerb and gully systems, although this remains the best-known and most popular option because of its cost-effectiveness with established design methods (see Figure 7.1). Other established options include drainage channels that are built at the edge of the road to receive runoff along their length (see Figure 7.2). These channels are available in various forms, including concrete surface water channels and prefabricated channel units covered either by gratings (grid units) or by lids with a longitudinal slot (slot units). Some systems incorporate a kerb in the drainage channel unit to provide a well-defined edge to the road and delineate footpaths (see Figure 7.3). Filter drains (see Section G14) consisting of a trench filled with gravel or other highly permeable medium are also widely used in road drainage. These are often referred to as French drains and could be categorised as a SUDS drainage system. Having been used extensively over many years for highway drainage, filter drains are one of the most established means of surface water drainage.

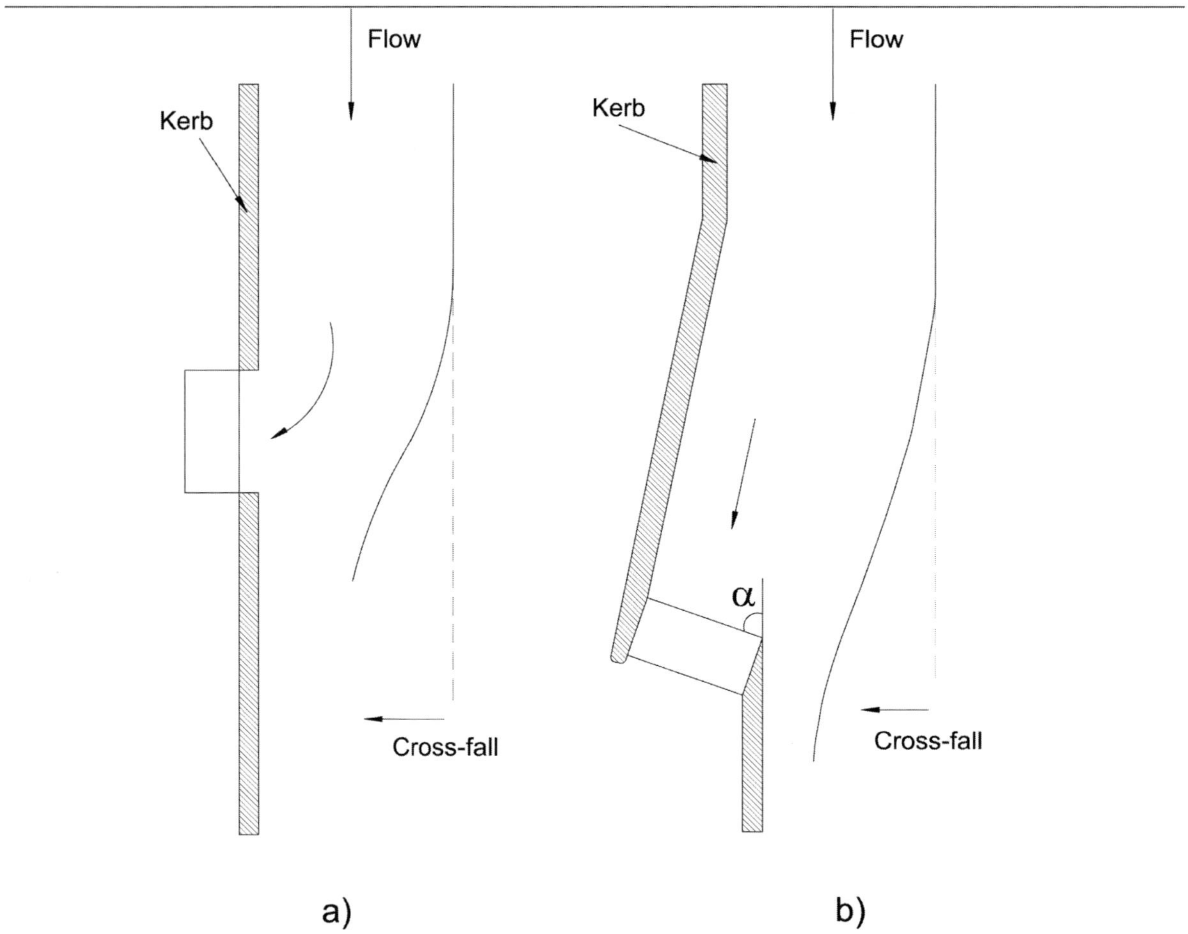

Plan layouts of kerb inlets
a) Straight
b) Angled (α between 50° and 55°)

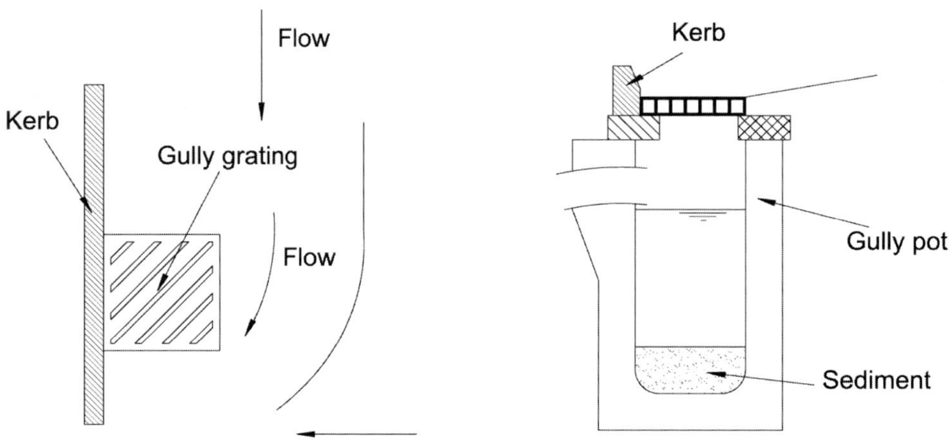

Schematic diagram of kerb and gully systems

Figure 7.1 *Kerb and gully systems – schematic diagrams*

Non-conventional systems aim to reduce the impact of the paved area on the natural conveyance and infiltration capacities of the site, maintaining the pre-development site hydrological behaviour. Permeable pavements are, in principle, not suitable for permanent road accesses (since they have limited ability to withstand traffic without significant deformation) and therefore should not be considered as an option. In some cases the road layout can be designed so that the surface water drains to adjacent permeable surfaces into which the water then infiltrates. Grass-lined shallow trenches (or swales), wetlands and ponds are among the various options for the drainage of roads, which can be used in combination with other methods to provide storage and pollution control. Soakaways can also provide a suitable means of drainage in permeable soils such as chalk. However there is some reticence among some engineers against this practice due to the long-term deterioration of performance from blinding caused by fine sediments and oils choking the structures.

7.1.2 Car park and pedestrian pavements

It is necessary to distinguish between impermeable pavements, which require additional elements of surface water drainage, and permeable pavements, which are drained by infiltration through the surface. This section concentrates on impermeable pavements since permeable pavements are covered in detail in Chapter 8.

Car parks and pedestrian precincts can be drained either by conventional or by SUDS methods and, unlike site roads, their low traffic loads and larger catchment areas render them more suitable for a wider range of options.

Conventional drainage options include: kerb and gully systems, terminal gullies at low points of the catchment, and prefabricated channel units covered by gratings (grid units) or by lids with a longitudinal slot (slot units) – see Figures 7.1 and 7.2. Some systems incorporate a kerb as well as the discharge channel in each unit to provide a fully terminated edge to the pavement and define footpaths (see Figure 7.3).

Sustainable drainage mechanisms offer a wide range of additional options, which include infiltration trenches and filter drains, permeable pavements, swales, ponds and wetlands (see Chapter 8 and Appendix G).

7.2 Standards

Normative and informative guidance on the design of paved areas is given by BS EN 752 *Drain and sewer systems outside buildings, particularly in Part 2 Performance requirements* (1997) and *Part 4 Hydraulic design and environmental considerations* (1998). The hydraulic design guidance contained in the Standard relating specifically to paved areas is presented in the form of informative annexes and was derived from the superseded BS 6367:1983 *Drainage of roofs and paved areas*.

Another European Standard, first drafted in 1994, which is in its final stages of preparation, will provide some information (but no hydraulic guidance) on channels used for pavement drainage: BS EN 1433:2002 *Drainage channels for vehicular and pedestrian areas*. Classification, design and testing requirements, marking and quality control. A normative document is also being developed that will deal with the hydraulic design of prefabricated drainage channels (linear drainage systems consisting of grated, slot or kerb units) to complement the above Standard.

Recommendations on design of surface water channels for drainage of roads are given in Advice Notes HA 37/97 *Hydraulic design of road-edge surface water channels* and HA 78/96 *Design of outfalls for surface water channels* (Highways Agency, 1997 and 1996).

7.3 Layout

7.3.1 General

General factors that influence or even dictate the options for the site layout were discussed in detail in Chapter 3; this section will concentrate on some specific issues. SUDS techniques, which are often combined with considerable landscaping work, will not be addressed here.

Conventional drainage systems for paved areas usually consist of the following elements:

- surface channels that collect the runoff from the paved area
- outlets from these channels (and associated chambers) that convey the flow to the below-ground drainage system.

The surface channels can be prominent features of the drainage system such as concrete surface water channels that are built along the edge of a road, or grated prefabricated channels. Alternatively, they may simply consist of a locally increased cross-fall in the pavement against the kerb to create a triangular channel.

In some cases, surface water channels used to collect the pavement runoff are not present and the outfalls (gullies) are placed at low points of the pavement.

7.3.2 Gradients

Gradients are introduced to allow quick drainage of the surface water and are also a safeguard against workmanship errors and pavement deformation that could otherwise result in local ponding. BS EN 752-4:1998 gives recommended gradients for three types of surface: "access roads", "paved areas" and "footpaths". No definition is given in the Standard for "paved areas", but it can be assumed that it includes car parks and pedestrian precincts as well as pavements used for purposes other than vehicular and pedestrian access. Table 7.1 shows the recommended gradients.

Table 7.1 *Recommended gradients for paved areas (from BS EN 752-4:1998)*

Location or type of gradient	Site roads	Car parks/pedestrian areas	Footpaths
Longitudinal gradient	1:15 max	–	–
Cross-fall or average camber	1:40 min	1:60 min	From 1:40 to 1:30
Against a kerb	1:100 min	1:100 min	–
Against channel units with kerbs	1:150 min	1:150 min	–
Super-elevation	1:25 max	–	–

7.3.3 Outlet spacing

This section deals with the spacing of outlets in conventional drainage systems such as kerb and gullies and drainage channels. In many positive drainage systems the spacing of outlets is dictated more by site layout constraints such as buildings than by hydraulic design considerations. However, it is possible to set out some general rules.

In kerb and gully systems the spacing between outlets depends on:

- the allowable width of flooding within the channel adjacent to the kerb
- the efficiency of the gully gratings
- the amount of flow that is allowed to bypass the grating.

Current design guidance is given in Contractor Report 2 (TRRL, 1984), but this publication has been replaced by Advice Note HA102/00 (Highways Agency, 2000). The latter is based on extensive experimental research conducted to determine a general design method applicable to any pattern of gully that conforms to BS EN 124:1994. The design method will take into account the local rainfall characteristics as well as the geometry of the grating and of the road. It should be noted that the guidance in BS EN 752-4:1998 has not yet been reliably validated for the UK.

The above publications do not cover symmetrical channels such as road-edge channels, for which specific design recommendations are given in Advice Note HA 78/96 *Design of outfalls for surface water channels* (Highways Agency, 1996).

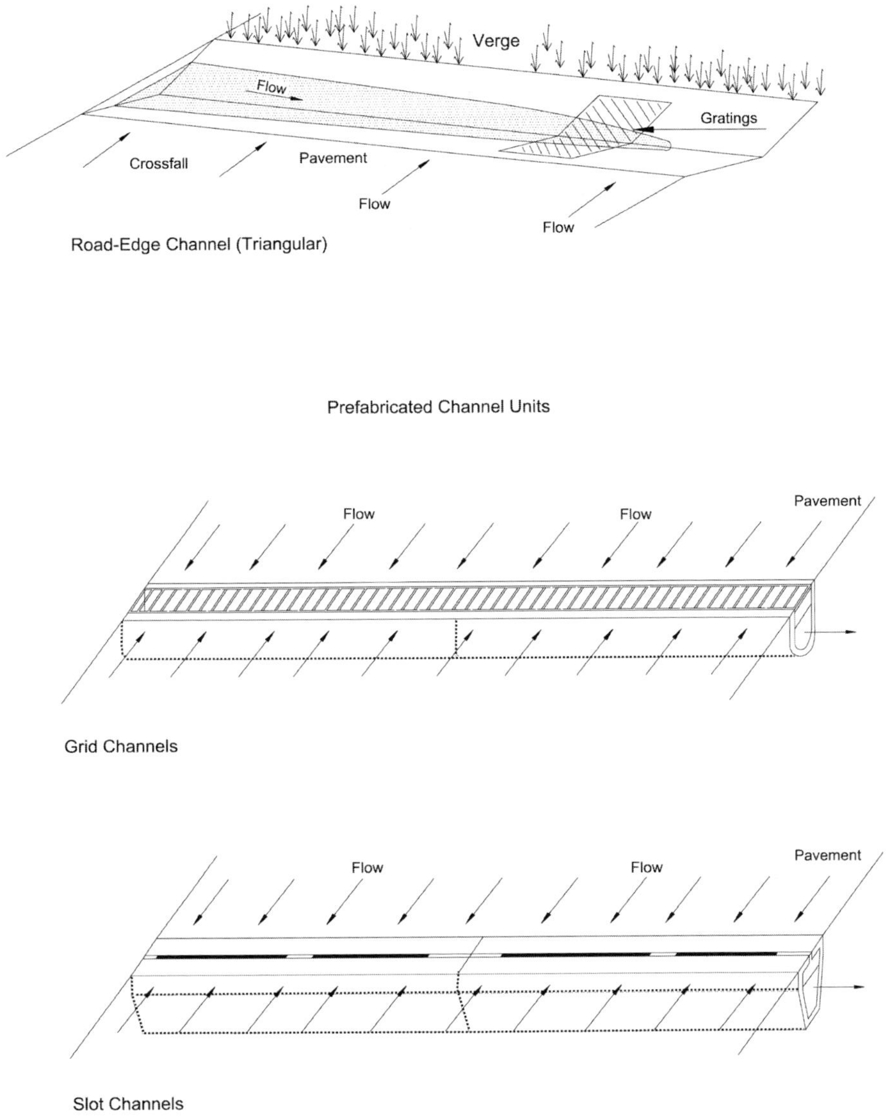

Figure 7.2 *Road-edge channels and prefabricated channel units – schematic diagrams*

Gully pots – see CIRIA Report 183 (Osborne *et al*, 1998) – are commonly used in the UK at entry points to drainage systems to prevent excessive amounts of sediment in the runoff entering the piped system. In an experimental study of a variety of gully pots, it was found that their through-flow capacity (ie the amount of flow that the gully pot can pass without becoming surcharged) varied between 11 l/s and 19 l/s. It was also established that, for each design of gully pot, the level of sediment retained inside the pot sump did not significantly affect the hydraulic capacity. The trapping efficiency of gully pots (ie the ratio of the sediment weight retained and the input sediment weight) depends on the sediment size and flow rate as well as on the retained sediment level inside the pot. Trapping efficiencies of 95–99 per cent are associated with medium sands (d_{50} = 0.8 mm) and of 40 per cent for fine sands (d_{50} = 0.12 mm).

Systems formed by prefabricated drainage units discharge directly into chambers when the channel flow capacity is reached. As explained in Section 7.5.2, the hydraulic principles behind the design of these systems are very different from the principles used for example to design kerb and gully systems. It is important therefore to use guidance that has been developed specifically for prefabricated channel units. Although manufacturers of these systems usually provide some flow capacity information, this guidance should be checked. To address the lack of guidance in this area, general design formulae have been recently developed at HR Wallingford (see Section 7.5.2) and their inclusion in British or European Standards is under consideration.

7.3.4 Large paved areas

The number and size of large paved areas has significantly increased in the UK in the last few decades mainly as a result of the expansion of out-of-town developments and park-and-ride schemes. These large areas can produce considerable amounts of surface water runoff in heavy storms, which need to be adequately quantified for the correct design of drainage systems. Current drainage design methods were developed for small areas and do not take into account factors that become significant when dealing with large paved areas. The main factors are:

- the geographical location of the site
- the duration and return period of the design storm
- the time required by the runoff to flow across the pavement to the point of discharge.

Research has recently been carried out to produce design guidance for two-dimensional catchment lengths from 10 m to 100 m, slopes from 1:150 to 1:50 and for four representative pavement surfaces (Escarameia *et al*, 2002). The design guidance in this document takes the geographical location into account and includes maximum water depths on pavements, corresponding peak flow rates and design critical storm durations in tabular form. It also includes values corresponding to 75 per cent of the maximum water depth and the length of time for which the water depths will remain above that value during the design storm.

7.3.5 Storage capacities

Some suppliers are developing innovative concepts to make use of potential storage volumes under paved areas and minimise adverse effects of impermeable areas such as higher peak flow rates and increased river floodwater levels. At the stage when the site layout is being defined, the entry and exit points need to be located so as to avoid low areas and thus enable the water to be temporarily stored during severe storms. The amount and frequency of water stored is a matter for detailed assessment depending on the particular conditions of the site and the use of the paved area. For example, it is possible to consider allowing a car park to flood for the duration of an infrequent storm to a depth of, say, 150 mm, and to smaller depths for more frequent storms. This design methodology may not be acceptable if the car park provides access to essential buildings such as hospitals and fire services where even temporary disruption is not acceptable, or if there is risk of flooding affecting the foundations of adjacent buildings.

7.4 Types of pavement drainage

An introduction to the options available for pavement drainage was made in Section 7.1 with some indication of the suitability of the two main types of application: access roads, and car parks and pedestrian areas. Each drainage type is described here in more detail. Chapter 8 and Appendix G should also be consulted for more information on permeable pavements, infiltration trenches, soakaways and wetlands.

7.4.1 Kerb and gully systems

Kerb and gully systems collect the runoff from triangular channels that are formed by cross-falls in impermeable pavements adjacent to the kerb (Figure 7.1). The flow is then discharged through gratings into gully pots or outfall chambers, which are usually provided with a sump for collection of sediment. Flow from the pot or chamber is conveyed to the sewer system by pipework. By retaining most of the coarse sediment washed off the pavement, this arrangement allows the sewer pipes to be built at fairly flat gradients. However, maintenance costs associated with the required emptying of the gully pots can be high, and the emptying operation, as well as the first heavy storm flush following a dry period, can trigger the concentrated release of pollutants into the sewer system. Kerb and gully systems are suitable for drainage of access roads as well as other paved areas. Gratings can also be incorporated in the kerb (kerb inlets parallel to the road or at an angle) and be combined with gratings installed in the pavement adjacent to the kerb (see Figure 7.3).

Gully gratings can also be used in isolation, ie without an adjacent kerb, as terminal outlets at the centre of local depressions in the pavement.

The Standard dealing with the specification of gully gratings is BS EN 124:1994 *Gully tops and manhole tops for vehicular and pedestrian areas*. The spacing of gratings and gully pots is discussed in Section 7.3.3.

Advantages

Kerb and gullies are a well-known means of positive drainage with established methods of design. Kerbs are necessary features in many cases to define footpaths and separate traffic areas from pedestrian areas.

Disadvantages

These systems are generally associated with relatively high capital and maintenance costs. There are concerns about the environmental impact of high concentrations of pollutants that are retained in gully pots and then released following heavy storms. Gully gratings over pots or chambers are hazardous for small animals by allowing their entry but preventing their escape. The recent movement to consider cyclists and providing the edge of the road as a dedicated cycle lane makes the use of gully gratings less acceptable.

The public may misuse the gully by tipping sump oil and floor washings into them.

Long-term level differences can result from the laying of additional wearing courses and subsidence due to poor compaction around the gully pot.

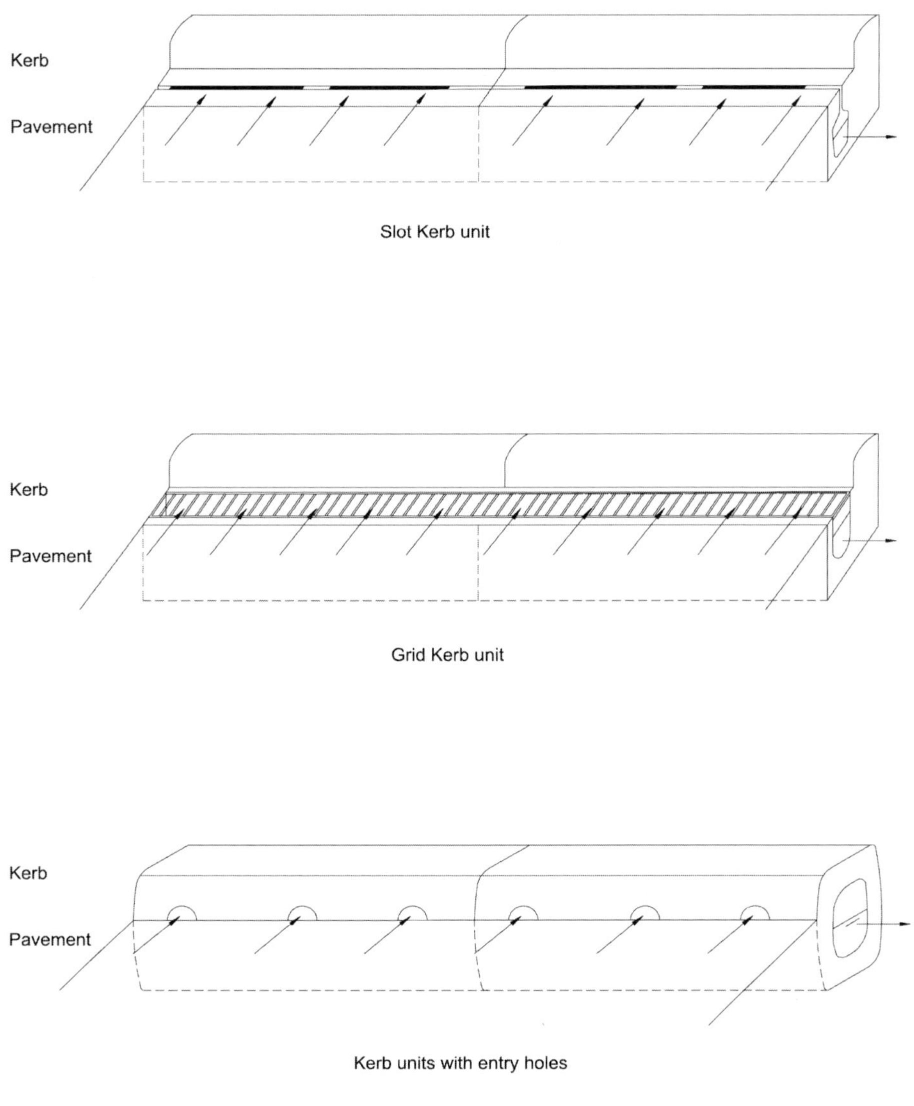

Figure 7.3 *Prefabricated kerb channels*

7.4.2 Drainage channels

Road-edge channels

Road-edge channels (also known as surface water channels) are usually slip-formed *in situ* in concrete but more environment-friendly alternatives involving vegetation lining are also being investigated. They can be triangular, trapezoidal or dish-shaped, with mild side slopes generally not exceeding 1:5 (Figure 7.2). These channels are usually built to follow the longitudinal slope of the road. Road-edge channels receive flow continuously along their length and are designed to discharge the flow through gratings into chambers when their capacity is reached. Because of their construction method (usually slip-forming, which requires specialised machinery) they become economically more viable for long stretches of road.

Prefabricated channel units

Channels formed by prefabricated units (also called linear drainage systems) are a common means of drainage in car parks, pedestrian precincts, roads and even in airports or inside buildings (see Figures 7.2 and 7.3). Like road-edge channels, they receive runoff from the pavement surface along their length and are designed to discharge into chambers when their capacity is reached. However, to increase their capacity, they are sometimes built with invert slopes that are bigger than the gradient of the adjacent pavement. Also they do not require the use of terminal gratings, as the flow is discharged directly into the outlet chamber. Prefabricated channel units are produced in various materials (eg concrete, steel and a range of plastic polymers) and shapes (from circular to polygonal shapes such as triangular or approximately rectangular).

No consensus exists on the terminology to adopt for the range of channel systems in the market, but it seems reasonable to use the categories defined in BS EN 1433:2002 *Drainage channels for vehicular and pedestrian areas. Classification, design and testing requirements, marking and evaluation of conformity*. This Standard defines the types of channel by the way through which the flows enters the conveyance channel. The categories are:

- grid channels (see Figure 7.2) – drainage channels with an open top, which is covered by metal or concrete gratings
- slot channels (see Figure 7.2) – drainage channel with a closed profile and a continuous or intermittent inlet slot on top
- kerb channels (see Figure 7.3).

Drainage channel with a kerb type profile and continuous or intermittent drainage openings. These openings may consist of holes in the kerb, or slots and grids on the channel adjacent to the kerb.

Advantages

The advantages are:

- these systems are open channels that are built flush with the pavement. Together with the fact that most systems have removable tops, this makes them easy to inspect and detect any problems. Some of the prefabricated units are also lightweight, which facilitates construction
- the use of channels means construction and flow depths are closer to the surface, one advantage of which is being able to discharge to a watercourse with high water levels
- the problem of new wearing courses and the difference in road level across a gully grating is avoided
- the tendency for local subsidence around gully pots is avoided.

Disadvantages

The disadvantages are:

- the capital cost of these channels can be high
- regular maintenance may be required to remove sediment that:
  - tends to deposit inside the channels particularly at the upstream end of a channel section where flow velocities are very low
  - gets trapped in the openings of the grids or slots thus blocking the ingress of water into the channels.

7.4.3 SUDS systems

Permeable pavements, swales, infiltration trenches, soakaways, filter drains, basins, wetlands and ponds are all facilities that are being used to drain road systems where appropriate. These are all detailed in Chapter 8 and Appendix G.

7.5 Design principles

As can be seen by the description of the types of system available given in Section 7.4, there are three main drainage groups that, due to their very different elements, require different design principles. These groups are:

- piped systems (eg kerb and gullies)
- drainage channels (eg grid units)
- SUDS (eg infiltration systems such as permeable pavements).

It should be noted that many systems categorised as "non-piped systems" often require pipe connections. Examples include linked soakaways or inlets into wetlands and ponds.

The hydraulic principles behind the design of the different types of system are presented below.

7.5.1 Piped systems

The hydraulic design of piped systems is discussed in Chapter 9.

7.5.2 Drainage channels

Drainage channels differ from piped systems essentially because they receive the runoff along their length (either continuously or at closely spaced intervals) and not at specific points of entry. By receiving flow in a continuous manner, the flow in drainage channels increases gradually until the capacity is reached and the channels need to discharge into outlet chambers. The design water depth in a drainage channel (ie the maximum allowable depth) can occur anywhere along the channel: at the upstream end of the drainage length for flat inverts to mid and terminal locations as the invert gradient becomes steeper. It is important to note that the channel capacity depends not only on the cross-sectional area and the channel material, but also on the invert slope and the rate of inflow into the channel.

At present there is no Standard covering the hydraulic design of drainage channels, particularly those formed by prefabricated channel units. Common forms of estimating the channel capacity include the use of Manning's and Colebrook-White equations (see Appendix F). However, these equations were not derived for spatially varied flows and therefore cannot easily be applied to the estimation of flow capacity. Design formulae and practical recommendations regarding the hydraulic capacity and also the effect of sediment in linear drainage systems have recently been developed (Escarameia *et al*, 2001). The capacity of the channels was found to depend primarily on the available cross-sectional area of the channel and on the channel slope. The length of the channel in relation to its design depth was also found to affect the capacity, and its effect depends also on the slope of the channel. The formulae are valid for channels at slopes not steeper than 1:30 and with length ratios (defined as the channel length divided by the design depth) not greater than 1000.

The particular case of drainage channels used for edge of road drainage (ie slip-formed surface water channels usually of triangular or trapezoidal cross-section), is adequately covered in Advice Notes HA 37/97 *Hydraulic design of road-edge surface water channels* and HA 78/96 *Design of outfalls for surface water channels* (Highways Agency, 1997 and 1996).

7.5.3 Sustainable drainage systems (SUDS)

For recommendations on the design of SUDS, see Chapter 8 and Appendix G.

8 SUSTAINABLE DRAINAGE SYSTEMS (SUDS)

Source control is a general, commonly accepted term covering all forms of drainage that limit the discharge of rainfall runoff at or close to its source. Other terms used are best management practices (BMPs) and sustainable drainage, although these are less descriptive of the drainage processes involved. SUDS is the term used throughout the UK water industry and is taken to mean best practice in the wider context of water quantity, water quality and the broader issues related to the control of rainfall runoff in the urban environment.

This chapter addresses all the various aspects of SUDS, and also comments on its current status in terms of its application by industry and the reason for their importance. SUDS encompasses a whole range of techniques; many of these have been around and accepted for many years whereas others are viewed with suspicion as their long-term effectiveness and value has still to be proven. This chapter introduces SUDS, but more detailed discussion is given in Appendix G, and specifically the following three aspects are considered for each SUDS component.

- description and purpose
- design principles
- maintenance requirements.

8.1 The SUDS philosophy

SUDS, source control, sustainable drainage and best management practices need to be defined clearly, as often these terms are employed to mean the same thing.

BMPs are the various methods and practices of drainage that are used to preserve nature's runoff characteristics and to ensure that the concept of environmental sustainability is most closely achieved. The objectives are therefore to:

- utilise rainwater as a resource (infiltration, environmental enhancement)
- minimise the volume of runoff
- minimise the rate of discharge
- minimise the pollution impact of runoff.

Source control refers to a subset of BMP techniques such as soakaways that aim to deal with rainfall at source. Management of flood flows farther downstream generally requires "hard" engineering solutions, which might be either avoidable or less expensive if source control methods were used.

Whether SUDS means the same as BMPs or whether it puts slightly more emphasis on wider issues such as the environment, amenity, ecology and other social aspects is not particularly important. The essential point is that the SUDS philosophy differs significantly from the traditional approach to drainage that has been practised for the past century or more.

A description of the SUDS components is given in the CIRA publications C521 and C522 (Martin *et al*, 2000a and b), which resulted from CIRIA Research Project 555 on sustainable drainage systems. These books are the SUDS design guides for Scotland and Northern Ireland (C521) and for England and Wales (C522). The differences between them are minimal and relate to variations in the regulators' roles in the two countries (see Chapter 2 and Appendix A). Further detailed design guidance is provided in CIRIA publication C609 *SUDS – hydraulic, structural and water quality advice* (Wilson *et al*, 2004), which provides comprehensive technical information on SUDS components.

such as Pratt, Ellis and Bettess regularly produce research material. Specific information on designing SUDS components is available in CIRIA C521, C522, C523 (Martin *et al*, 2000a, 2000b, 2001) and C609 (Wilson *et al*, 2004).

These guidelines for design of these structures have been partly derived from international research over the last 15 years. CIRIA is developing a comprehensive design manual on SUDS for the UK, which will take forward the work published in CIRIA C521 and C522. It is anticipated that this publication will be available in 2005.

8.3.2 Maintenance requirements

The responsibility for maintenance of SUDS systems varies depending on the type of SUDS technique used. Maintenance requirements for different SUDS techniques are discussed in the following sections. Assuming the organisation responsible is defined, there are two further aspects to be considered. The first is the degree of maintenance required; the second is that the drainage efficiency in the long-term service of each technique is often not certain. If it "fails" to be effective it can have a direct impact on another organisation or authority. This latter aspect is particularly the case in England, where sewerage undertakers have a duty to provide an effective drainage service. Drainage pipes might be sized to receive the runoff from a properly designed SUDS system, but if the SUDS system "fails" after several years' use, any resulting flooding could not then be addressed by the pipes downstream. Using a conventional design philosophy, the pipe drainage system would have been designed larger.

Local authorities and sewerage undertakers accept the SUDS techniques being promoted by the Environment Agency and SEPA represent good drainage management practice and are addressing some of these maintenance difficulties. Some parts of the system may be adopted as public sewers by the sewerage undertakers, but those parts that remain private (unadopted) must be designed with clear specifications for maintenance and responsibilities placed with the owner, management company or local authority.

In Scotland, a maintenance framework agreement was tried for two years. Local authorities and water authorities shared responsibility for drainage from land, roads and properties. The local authority maintained above-ground SUDS (such as swales and basins), while the water authorities maintained below-ground SUDS (filter drains etc). The cost of managing SUDS drainage was therefore shared. This form of agreement was found to have a number of limitations, which have since been addressed (see Section 8.3.3).

8.3.3 Legislative drawbacks

Under the current legislative framework SUDS are often considered to be landscape features, rather than an integral part of the drainage system. As a consequence, sewerage undertakers have no legal responsibilities for the operation and maintenance of many SUDS measures. In general, developers are reluctant to construct some SUDS options, because, without an adoption agreement with the sewerage undertaker, the maintenance of these units becomes the developer's responsibility.

The division of drainage responsibilities between several organisations is one of the principal drawbacks of the current legal requirements for drainage in England and Wales. Sewerage undertakers are legally responsible for the pipework and treatment of foul and surface drainage. Sewerage indicators consider it important to minimise operation and maintenance costs, with the result that even where they could adopt SUDS measures such as ponds, they often refuse to do so, preferring instead the low levels of operation and maintenance and predictable investment plans of traditional drainage infrastructure. The trial framework maintenance agreement tried in Scotland aimed to resolve this matter.

Various initiatives have taken place to address the issues of adoption and maintenance responsibility as a result of the difficulties caused by the current legislation. These difficulties for developing drainage systems had not been anticipated when the legislation was drafted. The initiatives are:

- the Water Bill in Scotland
- the SUDS Interim Code of Practice
- the production of model agreements for SUDS.

In Scotland a Water Bill has specifically addressed the issue of SUDS, and Scottish Water is defining those SUDS that will be adoptable. The document *Sewers for Scotland* is being modified to specify the design requirements of these SUDS units. The other SUDS units will either not be built or will be adopted by local highways authorities.

An Interim Code of Practice for SUDS was produced in 2004 by the National SUDS Working Group (NSWG). The group, chaired by the Environment Agency, produced a document to define the currenty status of SUDS and their application. Until legislation addresses some of the legal concerns of sewerage undertakers, this will be the key document for the use of SUDS in England and Wales.

CIRIA has produced a set of model agreements for the management of SUDS (publication C625, Shaffer *et al*, 2004). Although this has not been able to address many of the fundamental legal problems surrounding SUDS, it does improve the situation by providing templates for various types of contractual agreement and maintenance requirements for SUDS.

8.4 SUDS options

As stated earlier, there are several objectives in the use of SUDS. These can be divided into the two main categories:

- hydraulic control
- treatment of runoff to improve its water quality.

It should be emphasised that overall success of a site drainage system based on SUDS techniques depends on the integration of all the individual features of the catchment – a point highlighted in CIRIA publications C521, C522 and C609. The concept of total integration should also include possible reuse and recycling of surface water. The issue is not specifically addressed in this guide, but it is important to be aware that it can produce direct benefits of reduced flows downstream (and hence is effectively a SUDS technique) and can achieve resource efficiencies.

Most techniques provide some degree of hydraulic control, treatment and other benefits. Table 8.1 provides a comparison of the range of SUDS techniques together with the benefits they provide. However, it should be clearly understood that a pond designed to provide treatment is also likely to provide significant flow attenuation. Table 8.1 should be treated as indicative of their beneficial characteristics.

Table 8.1 *Qualitative comparison of SUDS techniques*

SUDS technique	Flow attenuation	Flow reduction	Water quality treatment	High visual amenity	Low maintenance required	Proven long-term reliability	Established design criteria
Water butts					✓	✓	
Sub-pavement storage	✓	✓ 1	✓		(✓)		
Swales	✓	✓ 2	✓	✓		✓	✓
Tank sewer storage	✓				✓		
Roof storage (flat)	✓				(✓)	✓	
Infiltration basin	✓	✓	✓	(✓)			
Paved surface flooding	✓					✓	✓
Detention basin	✓	✓ 2	✓	✓		✓	(✓)
Retention pond	✓	✓ 2	✓	✓		✓	(✓)
Permeable pavements	✓	✓ 1		(✓)			
Soakaways	✓	✓	✓		(✓)	(✓)	✓
Infiltration trench	✓	✓	✓		(✓)	(✓)	(✓)
Filter strip	✓	✓ 2	✓	✓			(✓)
Wetlands	✓	✓ 2	✓	✓			

Notes: (✓) it can be argued that it does fit this criterion
1 small reduction in volume if lined, high reduction in volume with infiltration
2 flow reduction only significant in smaller events

Each of these SUDS techniques will be briefly described to provide an understanding of where they might be applied and the constraints in using them. Design methodologies or rules of thumb are provided where they are known and the principles that should be used are given. Comment is also made as to maintenance requirements. The initial point of reference for all these techniques should initially be CIRIA publications C521 and C522 (Martin *et al*, 2000a and b), but CIRIA publication C609 (Wilson *et al*, 2004) provides more comprehensive guidance on the design and maintenance of SUDS. There are many other sources, particularly in the USA, that provide information on SUDS.

Two documents are:

- *Stormwater management devices: design guidelines manual* (Auckland Regional Council, 2002)
- *Stormwater management design manual* (New York State Department of Environmental Conservation, 2001).

8.5 Overview of CIRIA publications C521 and C522

The current principal reference point for SUDS design advice is CIRIA C521 and its parallel publication C522 (Martin *et al*, 2000a and b). They are much the same, one being targeted at Scotland and Northern Ireland, the other for England and Wales. This is to deal with the legislative differences in these regions rather than the climatic variations.

The documents give a structured approach to considering the process of drainage design using SUDS; choosing and integrating the various options. They discuss each option on the various issues of design, construction, operation and maintenance, amenity and safety. They then finish off with worked examples of the design of several SUDS options.

They do not provide many design criteria and reference is made to other documents that deal with design in greater detail. However, some SUDS units do not have well-established design procedures. A key design parameter, which is particular to the various types of SUDS ponds, is the concept of V_t, a storage volume calculation. V_t is the volume that is calculated as being equal to the runoff volume of 75–90 per cent of all storms occurring in the year.

The formula for V_t is:

$$V_t = 9\ D\ (SOIL/2 + (1 - SOIL/2).\ I)$$

where:

V_t is in m^3/ha

D is the M_560 rainfall depth from the Wallingford Procedure map (see Section 9.5)

SOIL is the WRAP soil value obtained from the Wallingford Procedure map

I is the impervious fraction of the area.

Where time series rainfall is available for the location of interest a check might usefully be made on the appropriate value of V_t.

The basis of using V_t (factored depending on the type of pond), is on experiential data. The climatic differences between UK and other parts of the world and even within UK will mean that the effectiveness of this parameter will differ according to location. In time, it is expected that more robust climate-specific criteria will be produced. As many of the benefits are related to biological processes that relate to light, temperature and fauna balance, it is believed that detailed design criteria for ponds cannot be entirely based on hydraulic/hydrology criteria.

Appendix G addresses the key hydraulic design and other issues that need to be considered when utilising SUDS components. Each SUDS component is considered in some detail.

9 PIPE DRAINAGE DESIGN

This chapter is primarily concerned with surface water drainage. Foul sewer design is covered briefly for completeness. Since the early years of the 20th century, the Rational Method has been employed for traditional drainage design, making use of rainfall intensity/duration/frequency information, and it remains in use today. More recently, the Wallingford Procedure proposed a minor modification of this method and is referred to as the Modified Rational Method.

This chapter explains the principles of pipe drainage design and describes the criteria normally used. It also clarifies what the Wallingford Procedure is, to allow engineers to understand the assumptions and parameters required when applying it. However, pipe sizing is now being considered as the first stage of a two-stage process. In recent years there has being a growing demand to be able to demonstrate the performance of the network under extreme event loading and the impact of flooding on the site. This analysis requires the use of simulation modelling. This chapter addresses:

- drainage systems
- design issues
- design criteria – foul systems
- design criteria – surface water systems
- the Wallingford Procedure and simulation modelling.

9.1 Drainage systems

There are three categories of drainage.

1. Separate foul and surface water systems.
2. Combined sewer systems.
3. Partially separate systems.

For the past 40–50 years most authorities have promoted separate foul and surface water systems on the basis that it is best to keep polluted water separate from "clean" surface water runoff. One advantage is that surface water system pipe sizes can be minimised by discharging to watercourses where the opportunity occurs. However, there is a sizeable minority of engineers who believe that this may not be good engineering practice. The problem with separate systems is that they are very rarely completely separate, as incorrect cross-connection always occurs by mistake or malpractice. The principal reasons for not having separate systems are:

- rainfall flows are significantly more per unit of area than foul flows; it does not therefore need many misconnections to result in flooding from foul sewers, which usually then leads to the construction of relief structures or combined sewer overflows (CSOs)
- the converse problem of connecting some foul systems to surface water drainage results in continuous low-level pollution of untreated sewage in streams
- it is now accepted that surface water runoff from urban development is not clean
- single-pipe systems account for approximately half the amount of pipework required and so should reduce construction and maintenance costs and misconnections.

Conventional wisdom is likely to continue with the use of separate systems for the foreseeable future. However, for developers the advent of SUDS has made this debate effectively redundant.

A **combined sewer system** is a single-pipe system that serves both the foul drainage and the surface water runoff. When designing such systems, care has to be taken in considering both the flows during dry weather as well as the much greater wet weather flows.

A **partially separate system** is traditionally one in which the foul sewer also serves some or all of the roof drainage while the roads are drained separately. This is not particularly desirable in terms of protecting the environment, as roof runoff is relatively clean compared to road runoff. As an alternative, the property wastewater can be connected to the road drainage while the roof runoff is discharged to soakaways where these can be built.

9.2 Design issues

In order to design a drainage system, many aspects need to be considered. Although this guide gives considerable detail about hydraulic design, there are some other aspects that need to be considered when carrying out initial design considerations for drainage pipework.

9.2.1 Rules of thumb for pipe sizing

Sizing of pipes is often carried out using programs based on the Rational Method. They are subsequently analysed using a hydrograph method to check for flooding performance. Approximate pipe sizes are quickly determined for small sites in the UK by a rule of thumb using a constant rainfall intensity of 35 mm/h. This value is often used, though adjustments are needed to take account of performance requirements or location. This approach should be used only to provide an initial assessment of the network, and should be followed by more detailed analysis to justify/modify the pipe sizes and gradients.

9.2.2 Selection of materials

Pipes are produced in a variety of materials, with the most common being vitrified clay, concrete, plastics and cast iron. The choice of material is governed by:

- overall installed cost – cost of pipe, bedding and installation
- maintenance – areas liable to blockage need to be able to withstand vigorous cleaning
- strength – to resist ground pressures and surcharge
- access – lightweight pipes for areas difficult to reach with machinery.

9.2.3 Site access

Traditionally, pipes are laid in straight lines with changes of direction taking place at access chambers. Chambers are also provided at changes of pipe size and changes of gradient as well as at intermediate points, all in order to provide access for routine cleaning and inspection, and for blockage removal.

For shallower depths, prefabricated inspection chambers are available, supplied with the pipe system used. Traditional brickwork chambers are used mainly where there is a need for a greater number of connections or where it would be difficult to install a proprietary chamber.

For deeper chambers, pre-cast concrete components are normally used.

The essential consideration in designing access is safety. If a chamber is intended for personnel access it must be possible to work safely and to affect a rescue if necessary; if it is not designated as suitable for personnel access, entry should be physically prevented.

Further details are given in BS EN 752-3.

9.2.4 Other site constraints

Pipework routeing and levels are often determined by site constraints, including topography (and the need to maintain cover to pipes while not entailing excessive excavation), other services and contaminated ground.

All site constraints should be identified as soon as possible in the design process including areas of site not available due to temporary works (eg scaffolding and site compound).

9.2.5 Services conflict

Foul and stormwater drainage are just two of the many services laid in the roads and footpaths in a new development. Care should be taken to ensure that services do not conflict. Most services are placed above the sewers, so there is little risk of conflict except where sewers have to be placed closer to the surface of the ground for topographical reasons.

The risk of conflict between the two sewers is generally avoided, as stormwater pipes are usually laid at flatter gradients than foul sewers. Foul systems are therefore normally deeper than stormwater networks. Where steep catchments dictate pipe gradients, it is again preferable to put foul sewers below storm pipes, as this both minimises pick-up of foul pollution by the stormwater system and minimises trench depths for the larger pipes, making it cheaper to construct.

It is considered good practice not to have the foul and surface sewer in the same trench above each other. This causes obvious problems at manholes. The use of a single, wide trench with the surface sewer offset, benched at a different level, tends to be more problematic to build than digging a separate trench.

9.2.6 Pipe bedding

The strength of the pipe and its bedding needs to be adequate to resist loading upon it, including loads from construction equipment. Pipes are normally laid on granular bedding. The bedding design is an important feature of ensuring adequate support for the pipe. Pipes that fracture are the main reason for excessive infiltration passing to treatment works.

9.2.7 Temporary drainage

The need for temporary site drainage during the construction phase should be considered. This can take the form of temporary ditches or use of the long-term storage system.

Settlement tanks or ponds may be required to prevent siltation of pipework systems, watercourses or infiltration systems. Poor control of silt during construction can cause premature failure of infiltration systems, as will washing off concrete and plaster into the drainage system. Consideration should be given to the installation of a temporary drainage system until the development is complete and vegetation has become established.

9.2.8 Adoption and approvals

The requirements of the sewerage undertaker and the Environment Agency, for adoption and consent to discharge respectively, should be established early on.

9.2.9 Safety

Drainage and sewerage systems are covered by the CDM Regulations, which cover safety during design, construction, operation, maintenance and demolition. Risk assessments at an early stage may rule out some options, even where they provide a least-cost solution. For example, no-dig installation may be required rather than very deep open-cut or, alternatively, pumping.

CIRIA publication C604 (Ove Arup and Gilbertson, 2004) on CDM and other guides on site safety can help designers understand their responsibilities and assess the risks related to design details. This does not absolve the engineer from taking responsibility for thinking through the risks for the proposed design and circumstances. Not only must the designer consider minimising risk, but the future contractor must also be informed by means of notes on those aspects that represent a significant level of risk in the construction process.

An important category of health and safety concern is the collapse of structures including trenches. There is a whole range of possible hazards, however, and thought should be given to the hazards that exist. Once the existence of a hazard has been established and the level of risk estimated, the following procedure should be carried out to minimise or remove the hazard.

1 Can the design be changed to avoid the risk?

2 Can the design be modified to reduce the risk – combat at source?

3 What controls can be applied to reduce the risk to an acceptable level (minimal or low)?

The resultant level of risk should be determined and categorised as Minimal, Low, Medium or High.

This process should be revisited at each stage of design.

9.3 Design criteria – foul systems

The principal documents that provide guidance on pipe design criteria are:

- *Sewers for adoption*, 5th edition (WRc, 2001)
- *The Wallingford Procedure vol 4 Modified Rational Method* (HR Wallingford and IH, 1981b)
- BS EN 752-4:1998 *Drain and sewer systems outside buildings. Hydraulic design and environmental considerations*
- BS EN 476:1998 *General requirements for components used in discharge pipes, drains and sewers for gravity systems*
- Building Regulations (England & Wales) 2000 and subsequent amendments
- CIRIA Report 141 *Design of sewers to control sediment problems* (Ackers *et al*, 1994).

9.3.1 Sewage discharged to a public sewer

Generally it is assumed that foul sewage will be discharged to a public sewer. In order for the drainage system to be adopted by the relevant authority the network needs to comply with *Sewers for adoption* (WRc, 2001). Table 9.1 summarises the guidelines provided by *Sewers for adoption* for the design of foul sewers.

Table 9.1 *Guideline foul sewer design criteria –* Sewers for adoption *(5th edition)*

Parameter	Foul sewers
Design method	Steady state
Minimum depth	1.2 m cover under highways, 0.9 m elsewhere
Maximum depth	–
Minimum sewer size	150 mm diameter for two or more properties
Design flow (pipe full)	Domestic: 4000 l/dwelling/day; industrial to be agreed
Velocity	0.75 m/s @ one-third design flow; 150 mm sewer at 1:150 for a minimum of 10 units is acceptable
Flooding	–
Roughness – ks	1.5 mm
Runoff	–

BS EN 752-4:1998 provides a set of criteria that is very similar. It suggests that 0.7 m/s should be achieved at least once a day, or that the gradient should be at least 1:diameter (mm). It emphasises the need to consider surcharge and flooding, pollution and protection against septicity. It specifically defers to national standards and guidelines for more detailed information.

Non-adoptable sewers

The criteria in Table 9.1 apply to adoptable sewers. Drainage within the property curtilage, and to the point of connection to the main sewer, is currently not adopted and is the responsibility of the property owner. Design guidance documentation for non-adopted sewers upstream of the adoptable system is defined by the Building Regulations.

The principal differences between non-adoptable sewers and adoptable systems are:

- foul sewers serving individual properties are usually 100 mm and laid a gradient of at least 1:60
- minimum cover to pipes can be reduced to 900 mm when under trafficked areas such as driveways, although this is highly dependent on the pipe material and vehicle loadings. In other circumstances cover is not specified but should generally not be less than 600 mm.

9.3.2 Sewage disposal if not discharged to a public sewer

This document is not intended to provided detailed information on sewage treatment, as in nearly all cases the developer will not be expecting to treat sewage flows.

Foul sewage is generally discharged into the sewerage undertaker's system. However, in remote locations it may be necessary to provide alternative means of disposal. These include:

- cesspools
- septic tanks
- packaged sewage treatment plants
- natural treatments (eg reedbeds).

Cesspools are simple storage devices that are emptied by tanker periodically. They should be watertight and are generally prefabricated plastic or concrete units. In England and Wales, their use is subject to Building Regulation approval and guidance is given in Approved Document H (for England & Wales), edition 2002. They are not allowed in Scotland and Northern Ireland. Technical guidance is also provided by BS 6297:1983.

The other forms of treatment all treat the sewage and dispose of the final effluent locally. A formal consent for the discharge is required from EA/SEPA, which also set the parameters to be achieved in terms of contamination. It should be noted that any of these systems is likely to be prohibited in aquifer protection zones, to protect public health.

Disposal of final effluent is often made to the ground using soakage trenches. Recent research has shown that biological films develop in the trenches that considerably reduce the effective permeability of the ground. Therefore, it is essential that the design of these trenches are not designed as for surface water trenches, but follows the design methodology given in BS 6297 (which is being revised to take account of the latest thinking). Local knowledge should be sought on the effectiveness of other foul soakaways in the vicinity.

Small sewage treatment facilities are particularly prone to failure due to grease in the incoming flows, originating from cooking and dishwashing. Effective methods of controlling grease are essential to meeting discharge standards. European Standards (BS EN 858-1:2002 and BS EN 1825-2:2002) dealing with grease separators are in the course of preparation, but in the meantime there is no definitive UK standard. Grease separators operate by providing a body of cool water, in which the grease solidifies. The low velocities in the separator allow the grease to float to the surface, where it is excluded from the outflow by means of baffles. The build-up of grease can be digested in the separator by means of biological additives or by periodically removing it by tanker. In either case, the separator should be sufficiently large to provide the necessary cooling.

Because of the increased organic and grease loading, food waste macerators should not be used in premises served by a small sewage treatment plant.

A European Standard is in the course of preparation, but in the meantime, technical guidance is provided in BS 6297. Design of sewage treatment facilities should normally be left to specialists, however.

Proper operation and maintenance are essential to achieve compliance with discharge standards and consideration to these items should be included in the design. Similarly, siting of the facility is crucial to avoiding nuisance.

Items to be considered include:

- access for tankers
- de-sludging
- routeing check on effluent quality
- emergency procedures
- operator training and holiday cover
- maintenance agreements
- provision of welfare facilities.

9.3.3 Drainage separators

Whether or not sewage is discharged to the public sewer, it is often necessary to consider using separators to deal with light oils, heavy oils or grease. As intimated in Section 9.3.2, present regulation is not clearly established, although work on all the relevant Standards is being carried out. The types of separators used in site drainage systems where appropriate are described below.

<u>In-situ oil/water separators</u>
These are normally formed in concrete and contain several separation chambers to retain the effluent for a sufficient time to allow the hydrocarbon content to rise to the surface under gravity for removal by skimming etc. This type of device is normally based on the American Petroleum Institute (API, 1969) design, or, for smaller units consisting of one to three chambers, on Property Services Agency Technical Guide 41 (PSA, 1987).

<u>Prefabricated oil/water separators</u>
These are normally constructed in glass-reinforced plastic (GRP) or polythene and also contain one or more chambers allowing hydrocarbons to rise by gravity and be stored at the surface prior to removal by pumping. This type of unit can contain a coalescing filter, which increases its efficiency and improves the quality of effluent discharged.

<u>Prefabricated grease separators</u>
Grease separators usually take the form of a rectangular chamber containing a number of baffles and having sufficient surface area and volume to allow grease to be trapped by gravity in the top of the unit. Removal is usually by hand at regular intervals. Some grease separators allow for the introduction of bacterial enzymes to convert the grease and protein matter into a water-soluble substance. The substance passes through the unit, thus increasing the time interval for grease removal by hand.

BS 8301:1985 *Building drainage* provides guidance on the use of these units, and will be replaced by the European Standards when they are issued. The Environment Agency also provides guidance in the form of a series of Pollution Prevention Guidelines. PPG3 *The use and design of oil separators in surface water drainage systems* (EA, 1998b) itemises sites that normally do and do not require oil separators, gives general design criteria and a method for calculating separator size based on six minutes' retention and catchment area.

The Environment Agency is drafting a new document entitled "Guidelines for oil separators", which will cover categories of use, test procedure and nominal size, and will describe a method of sizing separators based on the European test procedure. It will be applied to all Environment Agency regions.

9.3.4 Points of clarification on design criteria for foul sewers

The following points of clarification or comment on the various criteria are provided for information.

Minimum and maximum depth. It must always be remembered that pipe strength and pipe bedding define the load-bearing capacity of pipework. Specific design must be carried out where unusual loading takes place.

Minimum pipe size and gradient. It is generally accepted that 150 mm at a gradient of 1:150 provides self-cleansing conditions for more than 10 houses. The following explanation summarises the basis of this requirement.

1. The criteria for 0.75 m/s at one-third design flow allows flatter gradients and larger pipe sizes to be used and comes into effect for more than 500 properties.
2. It is not possible to provide 0.75 m/s at one-third design flow for fewer than 500 houses at a gradient of 1:150. However, the peak flow from a single toilet flush is approximately 2.3 l/s, which is the design flow for 150 houses. This flow enters the sewer at around 4.4 m/s.
3. BS 8005-1:1987 (superseded) and BS 8301:1985 suggest that any 150 mm pipe served by more than five toilets laid at a gradient steeper than 1:150 is acceptable.
4. A more detailed analysis of foul flows is available in BS 8005-1 clause 9.2, but most authorities interpret all the above as follows:

2–9 properties:	150 mm pipe at a minimum gradient of 1:60
10–270 properties:	150 mm pipe at a minimum gradient of 1:150
270–800 properties:	225 mm pipe at a minimum gradient of 1:150.

As domestic flows for small sewerage systems (fewer than 800 properties) are so low, the following is a useful guide. If connections occur at frequent intervals such that no part of the sewer is further than 30 m from a WC with a 9 litre flush then the pulses from intermittent flushes will ensure that normal sewage solids are transported. A design guide on the effect of water-saving low-volume flush systems (4–7 litre flushes) and other low-volume units has been produced by HR Wallingford (Escarameia, 2003) with respect to sewer gradient criteria. The research suggests that the self-cleansing criteria for gradients do not need to be changed, but that gradients of 1:150 for small numbers of properties above 10, as currently advised in *Sewers for adoption* 5th edition (WRc, 2001), are possibly most at risk from not achieving self-cleansing.

The minimum practical gradient for a 150 mm-diameter pipe is generally taken to be 1:150, so, based upon a design flow of (2000) l/house/day, this can serve up to 270 properties and a 225 mm pipe at 1:150 can serve 800 properties.

For more extensive sewerage systems self-cleansing is ensured by designing the sewers to achieve a cleansing velocity of 0.75 m/s at one-third design peak flow – bearing in mind that the pipe will generally be only part-full at this flow.

Peaking factor. BS 8005 (superseded by BS EN 752-4:1998) suggests a figure of 220 l/head/day, and peaking factors of 4 to 6 are used for foul sewer sizing to take account of diurnal effects. For combined or partially separate systems it advises 2.5 times DWF or 1500 l/dwelling.

Roughness. Guidance on the roughness of sewers for various materials is based on *Velocity equations for hydraulic pipes* (HR Wallingford, 1982). More detailed guidance on roughness may be found in *Tables for hydraulic design of pipes, sewers and channels, Volume 1,* 7th edition (HR Wallingford and Barr, 1998).

Roughness and velocity. CIRIA Report 141 *Design of sewers to control sediment problems* (Ackers *et al*, 1994) offers an alternative design approach based upon the solids-carrying capacity of flows in sewers. It is the state-of-the-art position on sediment transport in sewers. This is applied to both foul and surface water sewer design. It allows sewer gradients to be designed specifically for sediment loads for any pipe size. It is important to note that self-cleansing velocities increase with an increase in pipe size. It is possible that all pipe design will follow this methodology in the future. At the present, however, its complexity means that it is unlikely to be used by any but specialist engineers.

Dry-weather flow. The concept of dry weather flow is the base flow element of combined sewers, where base flow is the foul flow together with any other flow not generated by rainfall. Dry-weather flow has been defined for many years as PG + I + E. Its origins are uncertain. It is referred to in Circular 12/70 *Technical committee on storm overflows and the disposal of storm sewage* (Ministry of Housing and Local Government, 1970; issued in draft in 1963).

The definition of dry-weather flow is normally taken to be:

$$DWF = PG + I + E \tag{9.1}$$

where:

P is population

G is water consumption

I is infiltration

E is trade effluent

As all sewers leak to some degree it is important to have an allowance for infiltration. In practice, domestic water consumption is in the region of 165 l/head/day, though this varies a little with catchment area. In practice figures of 200 or 220 l/head/day are commonly used to take account of infiltration.

9.4 Design criteria – surface water systems

The principal documents that provide guidance on pipe design criteria for surface water systems are the same as for foul systems (see Section 9.3).

9.4.1 Adoption of surface water sewers

Surface water sewers are adopted if built to the standards and criteria stipulated in *Sewers for adoption*. It should be noted that the sewerage undertaker is only obliged to adopt sewers serving properties. Systems draining roads only come under the auspices of the local authority in its role of highway authority. Table 9.2 summarises the criteria defined in *Sewers for adoption* (WRc, 2001).

Table 9.2 *Guideline surface water design criteria –* Sewers for adoption *(5th edition)*

Parameter	Surface water sewers
Design method	The Wallingford Procedure, unless otherwise agreed
Minimum depth	1.2 m cover under highways, 0.9 m elsewhere
Maximum depth	–
Minimum sewer size	150 mm diameter
Design flow (pipe full)	1 year for slopes greater than 1 per cent 2 year for slopes less than 1 per cent 5 year where consequences of flooding are severe
Velocity	1 m/s @ pipe full
Flooding	Checks made for adequate protection* Should not flood for return period less than 30 years Simulation modelling is not required for sites less than 100 ha*
Roughness – ks	0.6 mm
Runoff	100 per cent from impermeable surfaces

* A check for adequate protection against flooding cannot be made without simulation. In practice nearly all systems are modelled to demonstrate that their performance is adequate for protection against flooding for extreme events.

9.4.2 Non-adoptable sewers

Drainage within the property curtilage is not adopted and is the responsibility of the property owner.

As is the case for foul drainage, the principal differences between local property sewers and adoptable systems are:

- sewers serving individual properties can be 100 mm. However, pipe sizes and gradients are based on pipe capacity and velocity, which are dictated by the same criteria as *Sewers for adoption*
- minimum cover to pipes can be reduced to 900 mm when under trafficked areas such as driveways, though this is very much a function of pipe material and vehicle loadings. Elsewhere, cover is not specified, but should not be less than 600 mm.

9.4.3 Points of clarification on design criteria for surface water systems

The following points of clarification or comment on the various criteria are provided for information.

Minimum and maximum depth. Pipe strengths and pipe bedding define the load-bearing capacity of pipework, so specific design must be carried out where unusual loading takes place.

Minimum pipe size and gradient. The concept of pipe-full design criterion is largely redundant in practice, as flooding is usually the controlling criterion. The use of pipe-full criterion helps guide the designer in achieving pipe sizes that are likely ensure this condition. Although it states that simulation modelling is not needed for sites of less than 100 ha, flooding can only be predicted using this method.

Pipes in surcharge are rarely watertight and can affect the surrounding bedding material, particularly in sandy soils. Care must therefore be taken to ensure that frequency of surcharge is not excessive in these circumstances.

Roughness. Guidance on the roughness of sewers for various materials is based on *Velocity equations for hydraulic design of pipes* (HR Wallingford, 1982). Fuller guidance on pipe roughness can be found in *Tables for hydraulic design of pipes, sewers and channels, Volume 1,* 7th edition (HR Wallingford and Barr, 1998).

Roughness and velocity. CIRIA Report 141 (Ackers *et al*, 1994) sets out an alternative design approach based upon the solids-carrying capacity of flows in sewers. It is the state-of-the-art position on sediment transport in sewers. This is applied to both foul and surface water sewer design. It allows sewer gradients to be designed specifically for sediment loads for any pipe size. It is important to note that self-cleansing velocities increase with an increase in pipe size. Although probably true, it shows that very large sewers require higher velocities (in excess of 3 m/s) to achieve total self-cleansing and that sewers should therefore be designed to allow a small amount of sediment deposition.

Formula A. This formula is unlikely ever to be applied to sewers in a new development. Formula A is applied to combined sewer systems as a criterion for storm flow relief. Details are added here for completeness because it is still used and referred to in national Standards.

Formula A was first promoted by the Ministry of Housing and Local Government (1970). This rapidly became an accepted standard due to its general ease of application.

Formula A is defined as:

$$Q = DWF + 1360P + 2E \tag{9.2}$$

where:

Q	is total flow in l/d
DWF	is dry weather flow in l/d
P	is population
E	is industrial effluent in l/d

This was arrived at by fairly non-scientific methods, in that the generally accepted approach of 6 DWF was thought to be generally appropriate though a slightly more rigorous standard should be set. It can be seen that 1360 is some eight times the current per capita consumption and that an allowance has also been added to dilute industrial effluent by a factor of three (from inspection of Formula 9.1).

The use of Formula A for setting overflows throttles is being (slowly) overtaken by the more recent Urban Pollution Management (UPM) procedure, the use of spill frequency analysis using time series rainfall and other methods. However, it is likely to be used as a general guide for many years to come as a useful and quick rule of thumb.

Runoff. There is an element of inconsistency in the runoff requirements in Sewers for adoption as the Wallingford Procedure is generally applied using the statistical UK percentage runoff (PR) equation, which does not apply 100 per cent runoff to paved surfaces and 0 per cent to permeable areas. Appendix E should be referred to for further information on the Wallingford Procedure runoff models.

9.5 The Wallingford Procedure and simulation modelling

The phrase "the Wallingford Procedure" is regularly encountered by those seeking consents for their proposed drainage systems. Its current meaning is not really very clear, which explains the confusion generally encountered. The following explanation describes the original derivation and meaning of the Wallingford Procedure and what it is generally accepted to mean now.

In 1981 HR Wallingford, with assistance from the Institute of Hydrology, completed a DoE-funded project by producing a document of five volumes and a range of software called the WASSP suite of programs. This was called "the Wallingford Procedure".

This suite of programs, which included a simulation program, is long obsolete. This simulation programme used hydrographic pipe routeing, but also took into account surcharge effects in the network and a rainfall-runoff model that was calibrated against recorded runoff information. This runoff equation has evolved over the years since 1981.

When authorities ask for the Wallingford procedure to be applied, this is now generally taken to mean the use of a simulation tool together with the UK calibrated runoff model. Current simulation software is effectively applying the same technique to network design and analysis and so is still viewed as providing "the Wallingford Procedure".

Appendix E provides more detailed information on the Wallingford Procedure, particularly the runoff models and the Modified Rational Method, which is still being applied for the design of small systems.

10 SITE STORAGE DESIGN – CURRENT AND PROPOSED METHODS

Developers have been finding it difficult to determine whether they need to provide storage to attenuate runoff and to establish the storage volumes that must be provided. This is partly because each site and receiving water is different and also because the various regions of the Environment Agency apply different criteria to the storage of stormwater runoff. At present, the regulator and planning authorities apply a range of criteria, depending on the region in which the development is to take place. This section provides guidance on the current status of storage design.

HR Wallingford has carried out research into this subject and has defined a possible national procedure for stormwater storage. The current approach generally applied by the Environment Agency is based only on a constant limiting discharge. This guide therefore provides information on both methods. Section 10.1 below summarises the current approach and the most commonly requested stipulations, and Section 10.2 presents the findings of HR Wallingford's research (Kellagher, 2002a and b) and the recommended approach.

Part A – Current regulatory requirements for storage (Section 10.1)

- regulatory basis for requiring site storage
- current requirements for storage design.

Part B – HR Wallingford's research findings and recommendations (Section 10.2)

- philosophy of approach for using site storage
- findings of DETR research on storage for developments
- proposed storage design criteria.

This is followed by a summary of methods for assessing runoff rates and storage volumes together with recommendations as to which are most appropriate depending on the circumstances.

10.1 Part A: current regulatory requirements for storage

10.1.1 Regulatory basis for requiring site storage

The involvement and responsibilities of the various planning authorities have already been detailed in Chapter 2. Nevertheless, the subject is worth revisiting with respect to stormwater storage requirements. Theoretically, local authorities are responsible for discharges to non-main rivers (ordinary watercourses), while the Environment Agency is responsible for designated rivers referred to as main rivers as well as for those designated "critical ordinary watercourses" (COWs). In practice, the Environment Agency is deferred to for advice on all planning applications with respect to river discharges. The Environment Agency is therefore not able to control stormwater runoff from new developments without the support of the local authority, which is able to stipulate runoff limits by means of planning agreements and conditions.

The Environment Agency comprises many regional offices and, although national guidance is provided, these act (at present) autonomously, which has resulted in a range of approaches to storage specification. Over the past few years, the requirements for discharge limits have tightened. The provision of storage is aimed at protecting rivers against spate flow effects and flood mitigation. This has led to the view that storage lower down the catchment is less desirable (to discharge ahead of the hydraulic wave coming down the river), and that stormwater discharge rates otherwise should be limited, particularly if known flood locations exist downstream.

Until now, the hydraulic criteria used by the Environment Agency have generally been to request a throttle to limit discharges to around 5–7 l/s/ha (although this can range from 1 l/s/ha to more than 10 l/s/ha). The upper limit is tending to come down, which demands greater temporary storage of stormwater runoff on site. The throttle value is arrived at by methods that range from the use of the *Flood studies report* to determine a site runoff rate, a selection of an arbitrary value, the use of the MAFF Report 345 runoff method, through to the calculation of the average unit rate of runoff of the catchment for a river discharge in flood. The return period requirements are usually for the 100-year event and using the site critical duration storm.

Research by HR Wallingford has shown that, for many locations, throttling and the provision of temporary storage makes little difference to the flood impact in the river, as the storm events that cause river flooding are usually of long duration with low average rainfall intensities (see Section 3.7.2). This is diametrically opposite to the present method of calculating site storage, as this is related to the event critical duration for the site and, in reality, results in very little of the storage volume being utilised during the events that cause river flooding. This does not discount the fact that rivers must also be protected from high-intensity (and usually spatially limited) events as well, which can temporarily cause high peak flows in the river locally, particularly where large developments discharge to small watercourses.

10.1.2 Current requirements for storage design

It must be stressed that the requirements for attenuation storage should always be considered in the context of the national emphasis being made for using SUDS to serve new developments. Storage is just one component of the SUDS techniques. In practice, until recently, attenuation storage has been the main mechanism used for protecting the receiving waters.

The general approach to attenuation storage is that it can only be avoided if the development area is particularly small because there is a minimum practicable vortex control unit or orifice control size. Occasionally, limiting the discharge from the site is not required if it is perceived that there is no implication for the receiving water. The position of the site in the catchment is usually an important aspect and is sometimes a reason for the Agency positively requesting that storage not be used. This is usually applied for developments that discharge into the lower reaches of rivers.

The approach to brownfield site development is often linked to current discharge characteristics rather than the equivalent greenfield area. The development of a brownfield site may be significantly more expensive because of the land treatment costs that are needed, so a flexible approach to stormwater runoff consent requirements is often taken.

Representations regarding surface water disposal are generally made in pre-draft local plan stage concerning flood defence (in line with DoE C30/92), conservation, landscape and water quality issues. PPG25 (DTLR, 2001) is a guidance document on development and flood risk and replaces C30/92 (DoE, 1992). PPG25 emphasises the need to use SUDS techniques for drainage of developments, and the Environment Agency has already included reference to SUDS in local plans in a number of regions.

The following is a non-exclusive list of the principal criteria generally used for specifying storage. This information was obtained from a nationwide questionnaire of Agency offices and developers, which was carried out as part of a DETR research project in 1999–2000. The last in the list is a recent document produced by HR Wallingford for the EA and Defra, which is intended to be applied nationally for all developments.

In nearly all cases the limit of discharge chosen is used with a 100-year rainfall critical duration event to determine the attenuation storage volume. In all instances, peak flow rate is the controlling criterion, and the volumetric aspects of the increased volumes of runoff caused by development are not explicitly addressed.

<u>Flood Studies Methods</u>
The *Flood studies report* (IH, 1975) methods and the supplementary studies carried out by the Institute of Hydrology (FSSR 1–17, 1975–1985) are recognised as having limitations for use on developments, as most sites are smaller than 50 ha, which is the officially recommended lower limit for using the FSR approach. Nevertheless, due to the limitations of other options, FSR is often used to determine the peak runoff from the greenfield site. Usually the peak flow for the return period of one year, or the mean annual flood is calculated and used as the peak discharge rate. Thames Region has introduced the concept of varying limits of discharge in line with runoff for any return period.

Although difficult to implement in practice, it results in low limits of discharge for frequent events and quite generous discharge rates for rare events. Although the use of FSR predictions of peak flow is not advised for catchments of less than 50 ha, the Institute of Hydrology method in its Report 124 (IH, 1994), when compared with MAFF Report 345 (MAFF, 1981), usually provides comparable values.

MAFF Report 345 Method

The MAFF Method is an alternative approach to determining peak runoff from a greenfield area. Its application to development sites is largely due to the 50 ha limit mentioned earlier. The advice given is that the MAFF Method should not be applied to areas larger than 30 ha. The basis of the MAFF work is on the measured runoff from fields and their subsoil drainage requirements for the farming community.

Total catchment analysis

There has been a move recently by the Environment Agency to analyse peak flow rates in rivers and, on a proportional basis, to determine the unit rate of discharge for the whole catchment. This figure – calculated by means of either FSR procedures or gauged flow records – is then used as the limit of discharge for the development site.

Arbitrary limits of discharge

The majority of Agency regions quoted maximum allowable runoff figures in the region of 5–10 l/s/ha. Higher values are sometimes used; an example being a value of 80 l/s/ha for a steep site in the north of England. The basis for choosing values is often indirectly linked to typical figures derived from the use of Flood Studies or other analytical methods.

Preliminary rainfall runoff management for developments

HR Wallingford was asked to provide an easy-to-use guide for site drainage requirements following on from research on the effectiveness of storage (Kellagher, 2002a and b). This implements the principles put forward in Section 10.2 of this guide and is intended for national application to eliminate the differences in approach taking place. The document, R&D Technical Report W5-074/A (2nd edn, Jul 2004), uses a look-up table approach incorporating graphics, figures and maps to enable preliminary quantification of site storage requirements and provides general guidance on drainage design.

For most regions, the majority of storage is implemented as oversized pipes or tanks, despite the Environment Agency's advocacy of open ponds for ecological and water quality reasons, wherever possible. Historically, sewerage undertakers have been unwilling to adopt (operate and maintain) such features.

The Environment Agency is now strongly promoting the use of SUDS techniques with the aim of mimicking greenfield behaviour in the rainfall runoff response from developments. The techniques allow volumetric runoff generation and the rate of runoff to be dealt with by means of various infiltration and storage techniques, some of which are relatively new while others are well established with a proven performance record. The proposed storage methodology described below fits closely with this policy objective and focuses on the need to determine which SUDS components provide long-term retention of stormwater runoff.

10.2 Part B: HR Wallingford's research findings and recommendations

HR Wallingford carried out a DETR research project to produce a "Guide to drainage storage assessment for greenfield sites" in 1999–2001, published as Reports SR 580 and SR 591 (Kellagher, 2002a and b). The work aimed to address the issue of the lack of a nationally agreed methodology, to test the effectiveness of current regulatory requirements and to determine an appropriate method for defining storage requirements.

The research project involved a steering group that included the Environment Agency,. The Environment Agency has taken forward the principles underlying the following recommendations in the production of a new guide (Environment Agency, 2004).

10.2.1 Philosophy of approach for using site storage

The philosophy of approach on drainage planning for greenfield sites is to use site storage to provide a mechanism by which the river regime can be maintained in its natural state, by minimising the differences between the developed and predevelopment catchment runoff. Storage is utilised to limit both the flow rate and volume of surface water runoff that drains directly off the site thus mimicking the predevelopment runoff state.

Although the proposed method does not explicitly use the word "SUDS" it is clear that some of the objectives are best met using SUDS techniques. The following section briefly explains the work and findings of the research carried out and the implications that it has with regard to storage design criteria.

10.2.2 Findings of DETR research on storage for developments

The study investigated the effectiveness of storage utilisation. This was carried out by selecting rain gauges near river hydrograph gauged data and assumed that a development site of 10 ha was located at that location. Analysis was carried out for 10 sites with a range of characteristics; the principal variables being soil type, size of river catchment, length of records and the spread of locations in the country.

The method of approach involved developing a simulation model for a theoretical 10 ha site and applying a range of throttles to limit runoff discharge, to measure the storage needed. The throttle rates used were 1, 3, 7 and 15 l/s/ha. This analysis was carried out using a variety of rainfall events. These were:

- events that occurred when the river was in flood
- all recorded rainfall events
- design storm events – critical duration of site
- design storm events – critical duration of catchment (T_p).

A second stage of this analysis then investigated the volume of runoff that passed to the river from the site and the amount that was retained on site during the flood event.

In addition, the study investigated and made recommendations on the various options for determining greenfield site peak runoff rates and runoff volumes. These were compared to urban drainage models for prediction of runoff volumes. A summary of the findings follows.

Design storage volume

A comparison was made of the storage requirements that were needed for a site by using standard design storm events with the recorded rainfall for that location. Using the standard approach of the critical duration event of a design storm, the storage volume predicted was closely reflected by the recorded data for the rainfall for the whole year for any given limit of discharge. This is exactly what should have been found.

The design storage requirements for rainfall events that took place during river flood events were correctly predicted by critical design events only for throttle rates of 1 and 3 l/s/ha. More generous throttles (7 and 15 l/s/ha) resulted in significantly less storage being needed than that predicted using the site critical design storm. This result is not unexpected, as the critical duration event for these higher limits of discharge is very short (one and two hours), whereas an event that is critical for a river catchment is often 12 hours or more. As throttles become tighter, the site critical duration event extends and is of a similar duration to that of the river.

Retention of runoff in storage tanks

This analysis checked on the validity of the assumption that the storage system was effective in reducing the flood impact of the urban runoff. A flood was considered as occurring when flow rates were above the 10 per cent frequency (Q_{10}) flow rate value. The analysis checked to see what proportion of the runoff volume passed to river by the time of the peak flow in the river and also the proportion that passed to the river by the end of the flood event.

The findings indicated that unless the throttles are very tight, in the region of 1 l/s/ha, the volume of runoff passing to the river is close to 100 per cent during the flood period and that there is rarely any significant volume retained on the site by the end of the flood event.

The implication of this result is that temporary storage using throttles either has to use very low limits of discharge or another method of retention is needed if the additional runoff generated from urbanised areas is to be mitigated. The concept of "permanent" or "longer-term" storage has therefore been introduced in the proposed criteria to cater for the additional runoff volume that urban development creates.

10.2.3 Proposed storage design criteria

If the basic premise is that greenfield conditions prior to development are to be replicated by post-development runoff, these research results imply that current methods of trying to protect the rivers are neither effective environmentally nor technically efficient. The following philosophy of approach proposed by HR Wallingford tries to take account of these findings. Although SUDS is not explicitly stated in the method, it is implicitly implied, as the concept of "long-term" storage requires reduction (infiltration) or longer-term retention of part of the volume of runoff, which might only be achievable using SUDS techniques.

The criteria have been broken down into six elements. This is not to add complication to the process of development, but is aimed at helping engineers to see that there are different aspects that need to be catered for in providing stormwater drainage. These six elements are a subdivision of two main objectives of serving the site and protecting the river. Use is made of the fact that different runoff rates take place from the greenfield site for different return periods. Therefore, to design and implement this concept to best effect, the following six elements should be considered. These criteria are listed in the likely order of frequency of return period.

1. **River water quality protection.** The aim is to protect the water quality of the receiving watercourse.
2. **River regime protection.** This criterion aims to protect against ecological and physical damage by minimising changes to the receiving watercourse's regime.
3. **Level of service protection.** The criterion is aimed at protecting the site from flooding from the drainage system.
4. **River flood protection.** This criterion ameliorates the risk of flooding in the receiving watercourse downstream of the development site.
5. **Site flood protection.** This criterion controls flooding of the site during extreme events.
6. **Catastrophic protection.** This is not explicitly a regulatory issue, but more a CDM requirement. It should be employed only where flooding of the site could lead to loss of life or when damage to property is likely to be excessive.

These criteria also subdivide into the two categories of site design and river protection. Although explicit consideration of all six elements should be made, there will be many situations where application of some, or even all, of the criteria may not be needed.

In practice, it is quite difficult to implement all these criteria just by using storage units and throttles to achieve the variable discharge limits to mimic greenfield response and the volumetric reduction or retention of "long-term" storage. The examples in Appendix D show how this procedure might be modelled using tanks and throttles. In practice, careful planning of the site together with the use of SUDS and a drainage model will be needed to implement all of these criteria. It must be stressed that pragmatic solutions may be needed that approximately achieve the objectives of the criteria required for the site development. The main principle behind the philosophy is that the developer should be entitled to discharge predevelopment runoff volumes at a rate equal to greenfield runoff from the site for any particular return period, but should retain additionally generated runoff volume for an extended period to limit the risk of exacerbating river flooding (see Section 10.2.1).

Table 10.1 summarises the criteria that are suggested for defining site storage for developments.

Table 10.1 *Suggested criteria for limiting greenfield site runoff*

Element	Return period (years)	Description
River water quality protection	Less than 1	• Minimum discharge rate 10 l/s • Interception of the first 5 mm of rainfall
River regime protection	1	• Minimum discharge rate 10 l/s • 1 in 1-year greenfield site discharge rate • 1 in 1-year site critical duration storm • No flooding on site
Site level of service protection	30	• Minimum discharge rate 10 l/s • 1 in 30-year greenfield site discharge rate • 1 in 30-year site critical duration storm • No flooding on site except where specific planned flooding is designed and approved
River flood protection	100	• Minimum discharge rate 10 l/s • "Long-term" flood storage for the additional development runoff volume compared to greenfield runoff • 1 in 100-year, 6-hour event
Site flood protection	100	• Minimum discharge rate 10 l/s • 1 in 100-year greenfield site discharge rate • 1 in 100-year site critical duration storm (storage) • 1 in 100-year 30 min high-intensity storm (flooding and flood routeing) • Temporary flood storage and controlled surface flood routeing
Catastrophic protection	200 or 500	• Risk to life or excessive property damage

It should be noted that the official Environment Agency position on climate change is to allow for a 20 per cent increase in peak river flows. It is suggested that a 10 per cent increase in rainfall is used in applying these criteria until research provides alternative advice. This reduction on the value of 20 per cent is proposed because (i) the storage implications of an increase in rainfall are significant and (ii) climate change predictions for extreme rainfall are very uncertain (UKWIR, 2004). Whether or not a 10 per cent allowance (uplift) should be applied to the derived values of greenfield runoff and volume is a decision related to providing a factor of safety. This ensures that climate change effects are formally being allowed for.

Each of the six criteria is briefly discussed to explain why they have been used. For explanation and discussion of the minimum discharge rate of 10 l/s, see Section 10.6.

River water quality protection

Rainfall on rural catchments initially has no runoff, except in very wet winter periods. Hence it is desirable to intercept this initial runoff "first flush" from a development site and prevent it reaching the receiving watercourse. This will not always be feasible, but it is an issue that should be considered and designed for, if possible. The objective of the water quality element is to minimise runoff of polluted water from development sites, thereby protecting the water quality of the receiving watercourse, especially during periods of low flow. Initial runoff from urban surfaces from rainfall is often highly polluted. Although 10 mm is sometimes stated to be a desirable interception volume, it is suggested that the first 5 mm is more achievable than trying to prevent larger rainfall depths from discharging directly to the river. Methods for dealing with this runoff usually involve the use infiltration, but care is needed in protecting the infiltration systems from clogging with sediments.

This criterion should be distinguished from V_t (treatment volume), as the idea of interception is to have no direct runoff, whereas a pond will discharge the same volume out as that coming in, subject to the water level being controlled by the outlet structure.

Where soil types prevent the use of infiltration, ponds designed to provide V_t (the treatment storage volume as defined in CIRIA C521 and C522 – Martin *et al*, 2000a and b), could be considered as effectively complying with this criteria. The value of V_t is derived using an empirical formula, but effectively aims to retain around 10–15 mm of rainfall.

The Environment Agency sometimes expresses a request for having an absolute minimum storage volume of 20 m^3 to intercept and trap accidental spills from tankers. Consideration of the risks of spillage for each development should be considered formally.

River regime protection
The objective of river regime protection is to minimise the impact of short response times and high flows of the runoff from a development site that discharge to a watercourse. This minimises erosion, high concentrations of suspended solids and other ecological damage. Return periods and limiting discharge rates should therefore be small and aim to cater for frequent rainfall events.

Site level of service protection
The developer should seek to ensure against flooding for a certain level of service and to comply with the philosophy of limiting runoff from the site to greenfield flow rates. This applies to both storage design as well as the drainage network performance. This is effectively the maximum storage volume that should be formally designed for without causing any inconvenience to the community. *Sewers for adoption* (WRc, 2001) uses 30 years as the level of service to be provided and this criterion would also seem appropriate for storage design.

River flood protection
Watercourses should be protected from additional volume of runoff taking place from developed areas compared to greenfield conditions for extreme wet weather periods. Analysis has shown (Kellagher, 2002a and b) that temporary flood storage using throttle rates above 2 l/s/ha have limited benefit in protecting a river with most of the runoff entering the river during the period of the flood (see Section 3.7.2). Effective discharge limits that reduce the volume of water to the watercourse during the flood period have to be very tight and this results in very large storage volumes being designed. The alternative approach suggested is to provide long-term storage for the additional rainfall runoff volume compared to that which would have taken place on the site before development. This storage would be "long-term" and be retained for disposal by infiltration or by using a limiting discharge of less than 2 l/s/ha.

This is not easily achieved using standard tanks and throttles, as rainfall intensities are generally very low for long-duration events when river flooding takes place. In practice, the best way to achieve this runoff volume control is to use ponds on a large regional basis, or certain SUDS components utilising infiltration on a site basis.

Site flood protection for extreme events
Sewers for adoption (WRc, 2001) defines the level of service to be provided to the community for drainage as 30 years. It also states that it is important to consider the implications of more extreme event. This also meets Environment Agency objectives of reducing the risk of flooding, usually stated at the 100-year return period. Compliance with this objective comprises three elements.

1 Protecting downstream communities from river flooding.

2 Consideration of the maximum flood levels in, or adjacent to, storage devices on sites to ensure housing or other important services are not affected.

3 Ensuring that flood flows across the site do not damage property.

The provision of "long-term" 1 in 100-year storage for the protection of the watercourse and the "end-of-pipe" storage volume to meet limiting discharge rates does not provide protection for the site from extreme thunderstorm events that overwhelm the development pipe drainage system. Flood routeing and the use of temporary flood storage in car parks or other areas needs to be properly designed. It is important to be aware that the network will be heavily surcharged, so flooding will take place at many points across the site and consideration will need to be given to the impact of flood flows and their routeing.

Where "long-term" storage is provided for sites with sandier soils, calculations will generally show that this dominates the storage needs.

Catastrophic protection

The possibility of extraordinary rainfall and flooding and its implications should always be considered. It is rare that designing for such return periods can be justified on economic grounds, but where there is a risk that a life-threatening situation could develop, it is important to demonstrate that these circumstances have been duly considered. For instance, where a river or dam might breach its embankments, or a large water main burst, and inundate an urban area, investigation of the extent and implications of such an event would be justified. Particular emphasis should be placed on topography and the routeing of flood flows to prevent loss of life. The Environment Agency and the insurance industry regularly use return period criteria of 200 and 500 years.

Figure 10.1 illustrates the various aspects of storage design for a site.

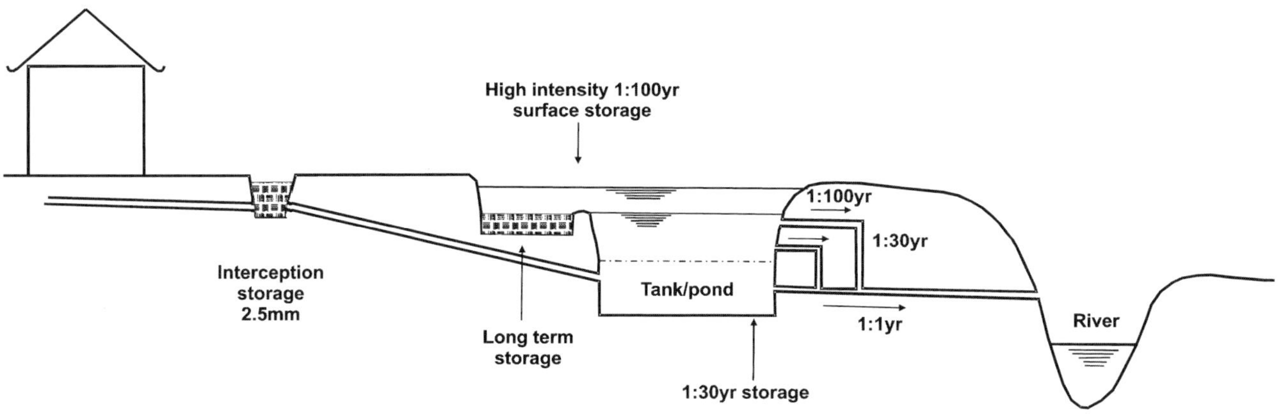

Figure 10.1 *Schematic of site storage design*

Two examples are used to illustrate both the existing and the proposed new method of approach. These are given in Appendix D.

10.3 Assessment of greenfield site rate of runoff

Numerous hydrological techniques are used for estimating greenfield runoff rates. Although briefly alluded to earlier in this chapter, this section reviews the techniques commonly used by the Environment Agency, developers and consultants in calculating greenfield runoff rates from sites that are to be developed. A more detailed discussion on this subject can be found in SR 580 and SR 591 from HR Wallingford (Kellagher, 2002a and b). The methods include:

- MAFF Report 345 (MAFF, 1981)
- techniques based on the Rational Method
- TRRL LR565 *The estimation of flood flows from natural catchments* (Prudhoe and Young, 1973)
- *Flood studies report* statistical and rainfall runoff methods (Institute of Hydrology, 1975)
- IH Report 124 *Flood estimation for small catchments* (Institute of Hydrology, 1994)
- *Flood estimation handbook* (CEH, 1999).

The two principal methods that engineers employ for determining peak flow rates are MAFF Report 345 and IH Report 124. The different parameters used make it difficult to carry out a direct comparison between these two methods, but for typical catchments the difference is relatively small. The other methods are not covered in this guide. Comment on the implications of the *Flood estimation handbook* is also given, as this document is of national importance and is likely to become a cornerstone for hydrology in UK in the coming years.

The methods detailed in the *Flood studies report* and the *Flood estimation handbook* (FEH) are applicable to a wide range of catchment areas ranging from some 0.5 km^2 to 5000 km^2. The MAFF Report 345 Method should only be applied to estimate the flow from small, rural, ungauged catchments (ie generally catchments of 30 ha or less). It should be noted that predictions of runoff and flows for small catchments are not going to be very accurate and this particularly applies to small greenfield sites for development.

The choice of method of assessing greenfield runoff is important, because the predicted peak rate of runoff and volume of runoff are used to determine the runoff control measures for the developed catchment.

10.3.1 MAFF Report 345 technique

Background to the MAFF Method

MAFF Report 345 details a technique that is primarily focused on providing information to determine the size of pipes required for field drainage systems. The method is based on measurements taken several small, rural catchments.

The equation to estimate runoff from a site is of the form:

$$Q = S_T F A$$

where:

- Q is the peak flow in l/s
- S_T is the soil type factor that ranges between 0.1 for a very permeable soil to 1.3 for an impermeable soil
- F is a factor that is a function of the following catchment characteristics: average slope, maximum drainage length, average annual rainfall. The F number can be estimated from a nomograph included in the MAFF report
- A is the area of the catchment being drained in hectares.

Guidance on the values of the above variables is given in the MAFF report, together with a nomograph, which can be used to estimate the flow. It is advised that this method should not be used for catchments exceeding 30 ha (this catchment size is given as a limit in the report). The predicted peak flow resulting from the MAFF equation should be taken as being the one-year return period flood and not the mean annual flood for the catchment. Flow rates for higher return periods can be calculated using the appropriate *Flood studies report* regional growth curve.

Advantages and disadvantages of the MAFF Method

The MAFF Report 345 method for calculating flows:

- provides an easily applied cheap, simple and quick method for calculating peak flows from small, rural catchments
- requires relatively few variables, which have to be estimated from Ordnance Survey mapping
- regulatory authorities are tending to apply/accept this method of analysis.

The disadvantages of the method are that:

- the method should not be applied to catchments more than 30 ha.

10.3.2 Flood estimation for small catchments (IH Report 124)

Institute of Hydrology Report 124 was published in 1994 and describes research for flood estimation for small catchments. The research was based on 71 small rural catchments. These catchments are not small relative to typical developments, however, as they are defined as having areas less than 25 km^2. A new regression equation was produced to calculate $QBAR_{rural}$ the mean annual flood for small rural catchments.

$QBAR_{rural}$ is estimated from the three variable equations shown below:

$$QBAR_{rural} = 0.00108 AREA^{0.89} SAAR^{1.17} SOIL^{2.17}$$

where:

AREA is the area of the catchment in km^2

SAAR is the standard average annual rainfall for the period 1941–1970 in mm

SOIL is the soil index, which is a composite index determined from soil survey maps that accompany the *Flood studies report.*

This equation should be used in preference to the equivalent equation detailed in *Flood studies supplementary report* no 6 for use on small rural catchments. The equation can also be used as an alternative to the six variable equations in the *Flood studies report.*

The QBAR can be factored by the UK *Flood studies report* regional growth curves to produce peak flood flows for a number of return periods.

<u>Advantages and disadvantages</u>
The advantages of the IH Report 124 Method are:

- the research carried out to produce the three variable equation was based on small, rural catchments
- the research was based on 71 small, rural catchments; significantly more than either the MAFF or TRRL research
- peak flows for various return periods can easily be calculated by applying the *Flood studies report* regional growth curves
- the method is simple and relatively easy to use.

The method has the following disadvantages:

- in theory the method should be applied only to a catchment drained by a well-defined watercourse, not to a small greenfield site
- it should not be applied to catchments that are smaller than 50 ha
- it does not have a catchment slope function.

In practice, comparisons with the MAFF model for smaller areas do not indicate that the model is problematic for catchments of less than 50 ha. However, if used in these circumstances, a comparison with the MAFF method is advised.

10.3.3 Flood estimation handbook

In 1999 the Centre for Ecology and Hydrology (CEH) produced the *Flood estimation handbook* (FEH). The methods of flood estimation detailed in the FEH unofficially supersede those in the *Flood studies report* (FSR – IH, 1975) and the *Flood studies supplementary reports* nos 1–17 (IH, 1975–85) as standard practice in the UK for assessing river behaviour. The Environment Agency has formally stated that the FEH should be used in preference to FSR techniques.

The FEH research carried out an enormous amount of data processing, not only on river flow and related catchment characteristics, but also rainfall analysis. Both these aspects are discussed briefly.

River flow estimation
The *Flood estimation handbook* provides two main approaches for flood frequency estimation. These are:

- statistical methods
- *Flood studies report* rainfall-runoff method.

The *Flood estimation handbook* is supported by three software packages:

- WINFAP-FEH to support the statistical procedures for flood frequency estimation
- Micro-FSR to apply the *Flood studies report* rainfall-runoff method
- FEH CD-ROM, which presents catchment descriptors for four million UK sites and implementing the rainfall frequency estimation procedure.

For the estimation of runoff from greenfield sites the Micro-FSR package and FEH CD-ROM are applicable. The FEH CD-ROM includes a digital terrain model of the whole of the UK produced by the Institute of Hydrology from 1:50 000-scale maps. The CD-ROM allows the delineation of catchment boundaries to be carried out automatically. The catchment boundaries based on the terrain map are inevitably approximate. However, it should be noted that discrepancies are most likely to arise for small catchments.

The general philosophy behind flood frequency estimation in the FEH is as follows.

1. Flood frequency is best estimated from gauged data.
2. Where gauged data are not available data transfers from a nearby or similar catchment are useful.
3. Estimation of floods from catchment descriptors alone should be used as a method of last resort.

The FEH provides catchment descriptors for all sites draining an area of 0.5 km^2 (50 ha) or greater. The lower limit indicates that:

- very small catchments are poorly represented in the data sets used to calibrate the models for estimating flood frequency from catchment descriptors
- digital terrain and thematic data may not be well resolved on very small catchments
- the digital approach and the complexity of the tool means that greenfield runoff estimation for small sites makes it very difficult to apply, both for small development sites and by any else other than hydrologists.

Advantages and disadvantages

The *Flood estimation handbook*'s advantages are that:

- it uses pooling and transfer of flow data to estimate flows from ungauged catchments
- it calculates the catchment area and descriptors digitally from a digital terrain model
- the pooling of the flood flow data for defining the growth curve is flexible and tailored to fit the subject site.

However, the *Flood estimation handbook*'s drawbacks are that:

- it requires detailed hydrological knowledge to apply the techniques correctly
- it requires detailed knowledge of the use of the FEH CD-ROM and WINFAP_FEH software
- catchments below 0.5 km^2 cannot be defined on the FEH digital terrain model
- digital catchment areas are predefined.

Rainfall analysis

The FEH rainfall analysis was carried out for the period 1971–1990 and had approximately four times as much information as that available for the FSR analysis, which was based on data between 1941 and 1970. The results, whether affected by climate change, the larger data set or more advanced processing methods, show significant differences to the FSR rainfall characteristics. These differences reach 40 per cent in parts of the south-east of the UK, while other areas have relatively similar characteristics. These differences nearly all show a greater rainfall in the FEH assessment of rainfall depth. It is therefore very important to take account of the FEH rainfall information, even if other techniques are being used to assess flow rates. This applies, of course, not only to greenfield analysis but also to development runoff analysis.

10.4 Calculation of greenfield runoff volume

Although there is little stipulation currently to take account of the volume difference between greenfield runoff and development runoff, it is becoming clear that this is an important issue. As with the estimation of peak flow rates, the science behind the estimation of runoff volume is of limited accuracy.

The greenfield runoff volume for a particular rainfall depth can be estimated for extreme events by calculating the percentage runoff from the site. The volume of the runoff per unit area from greenfield areas can be calculated from the product of the rainfall depth and the percentage runoff. The percentage runoff from a greenfield site can be calculated using the technique detailed in *Flood studies supplementary report* no 16. The percentage runoff for a rural area (PR_{RURAL}) is given by the following equation:

$$PR_{RURAL} = SPR + DPR_{CWI} + DPR_{RAIN}$$

where:

SPR is the standard percentage runoff, which is a function of the five soil classes S_1 to S_5

$$SPR = 10S_1 + 30S_2 + 37S_3 + 47S_4 + 53S_5$$

DPR_{CWI} is a dynamic component of the percentage runoff. This parameter reflects the increase in percentage runoff with catchment wetness. The catchment wetness index (CWI) is a function of the average annual rainfall. The relationship is shown in Figure 10.2.

$$DPR_{CWI} = 0.25\ (CWI - 125).$$

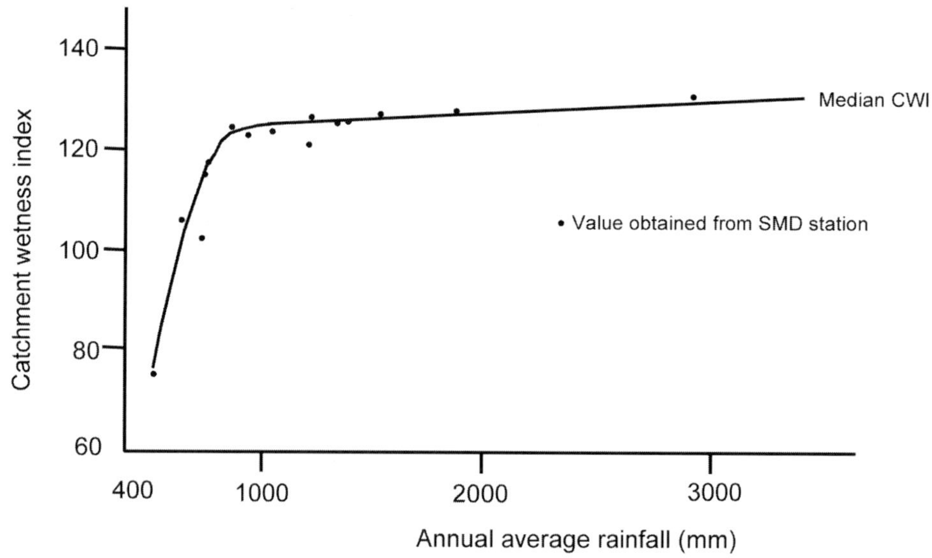

Figure 10.2 *CWI* vs *SAAR* – Flood studies report

The DPR_{RAIN} is the second dynamic component that increases the percentage runoff from large rainfall events.

$$DPR_{RAIN} = 0.45(P - 40)^{0.7} \text{ for } P > 40 \text{ mm}$$

$$DPR_{RAIN} = 0 \text{ for } P \leq 40 \text{ mm}$$

where:

P is the depth of rain that falls during the design storm.

The FEH recommends that the estimate of percentage runoff for a storm of P mm should be checked against the runoff values in neighbouring gauged catchments. Figure 10.3 indicates the relationship between runoff per unit area and rainfall depth for mean annual rainfalls of 1000 mm for the five soil types used in the *Flood studies report.*

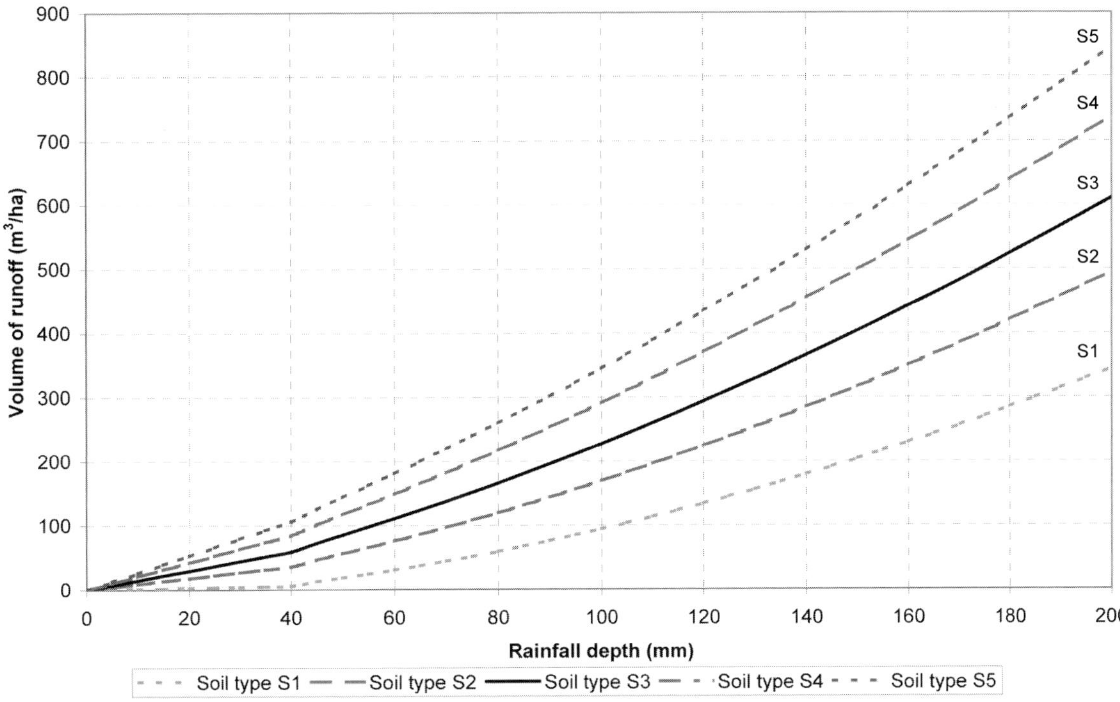

Figure 10.3 *Volume of runoff versus rainfall depth for a mean annual rainfall equal to or greater than 1000 mm*

The derivation of this equation is for extreme events and for catchments that are larger than those of development sites. Its accuracy is therefore to be treated with great caution. However, if account is to be taken of the volumetric effects of development, this is one of the scientifically based methods for assessing runoff volumes. It has the advantage of simplicity, so a rapid assessment of the impact of development can be made with respect to runoff.

The key feature of this formula is the important influence of soil (and SOIL index). In practice, it indicates that developments on sandy soils create massive additional runoff, but developments on clays do not. This is perhaps obvious, but it has significant implications for development costs in these areas. Other parameters have very little influence.

It should be noted that there is no allowance for slope of the catchment, which intuitively one might expect there to be. At present this approach is considered to be best available practice, however.

10.5 Assessment of development runoff and storage volume

10.5.1 Rate of runoff

Runoff from positively drained paved areas is effectively instantaneous by comparison with greenfield runoff. The runoff rate therefore reflects the intensity of rainfall with a little attenuation being provided by the filling of depression storage, surface runoff routeing and pipe routeing. This is true for all rainfall up to around 20–30-year return period events, when short-duration thunderstorms have intensities that are so great that brief temporary flooding takes place because the capacity of the pipe system is inadequate to cope with the volume of water.

From a site storage point of view, it is therefore relatively unimportant to determine peak flow rates from the site except to be aware that it is rapid. Information relating to the routeing processes of rainfall runoff used in models is available in the *Wallingford Procedure* (HR Wallingford and IH, 1981) or *Wallingford Procedure for Europe* (Kellagher, 2000). The important feature in determining any storage volume is the percentage rainfall that directly drains and runs off into the drainage system, and the shape of the storm profile being used.

10.5.2 Runoff volumes

The determination of storage is a function of runoff volume and the critical duration needed to fill it, which effectively approximates to the duration of the rainfall. The Wallingford Procedure runoff models allow this storage to be determined as it is a volumetric method using rainfall hyetographs. The Wallingford Procedure runoff equations are given in Appendix E.

Rainfall profiles used in UK are designed as either summer or winter profiles. These are symmetric with the same volume, but with the maximum winter intensities being a little less than the summer design events. Comparison of the FEH with FSR has shown that the rainfall events have volumes and intensities that can differ significantly for many locations in UK.

The critical duration event for any specific limiting discharge may be either a winter or summer profile depending on the relative volumes above the effective runoff intensity threshold for that event. The more restricted the limiting discharge, the longer will be the critical duration event. Figure 10.4 illustrates the effect of rainfall profile and limiting discharge in the determination of storage volumes.

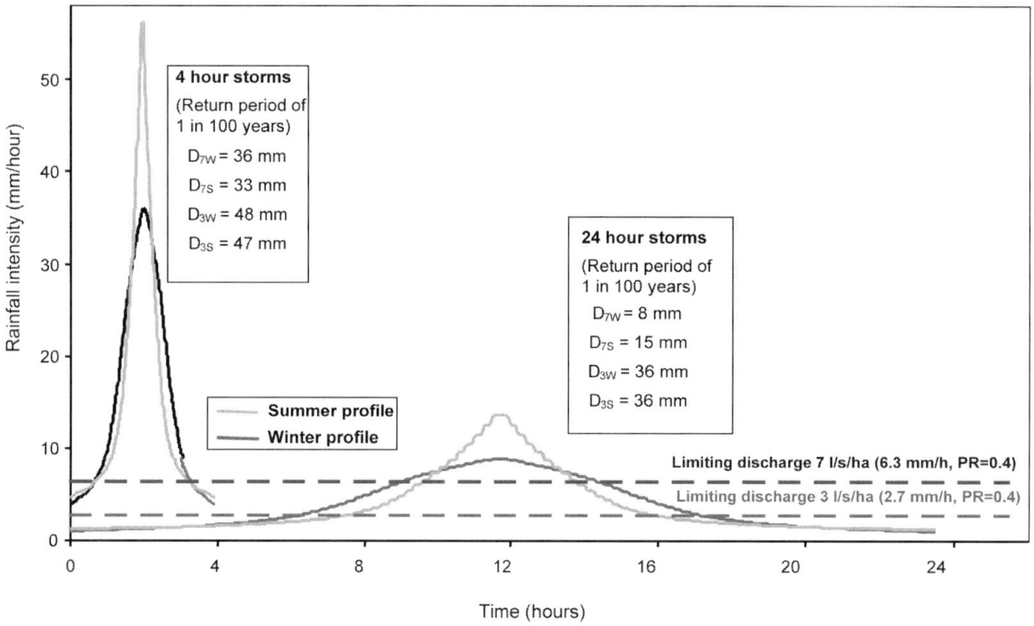

Note

D_{7W} refers to the effective depth of storage design in terms of rainfall depth for a winter profile for a throttle of 7 l/s/ha

Figure 10.4 *Rainfall profile effects on critical duration events*

It can be seen from Figure 10.4 that the choice of event duration is critical in determining the storage volume. Currently, the worst event is selected (usually around two to four hours), although, as stated earlier, these events rarely relate to the rainfall that causes rivers to flood. Figure 10.4 is based on the assumption that rainfall can be calculated as an equivalent runoff for a theoretical catchment with a total percentage runoff of 40 per cent with two throttle rates. This helps to illustrate the effect of choice of event profile and event duration.

Comparison of the urban runoff models (Wallingford Procedure) and the runoff calculation method of FSSR 16 can show that a SOIL of type 4 (clay) actually has slightly more runoff than post-development runoff volumes at the 100-year return period, although at lower return periods less runoff is predicted. This is intuitively incorrect, as paved surfaces have around 80 per cent runoff whereas the FSR analysis cannot give values much greater than around 55 per cent. The reason for this apparent discrepancy is that the Wallingford Procedure runoff assumptions for the permeable surfaces of the development catchment are lower than the FSSR 16 values. It can be argued that this is correct due to landscaping and other development effects, but these issues should be specifically considered. Care should therefore be taken in applying this procedure using these two equations for determining runoff volumes.

An alternative, more conservative, method for assessing the additional runoff volume assumes that pervious areas continue to generate the same runoff after development as before and a constant runoff of 80 per cent takes place from paved surfaces. An equation can therefore be developed that treats the paved area as having runoff of 80 per cent, and pervious areas as having runoff characteristics that are the same as those prior to development. This method might be termed the SPR fixed percentage approach. It can be modified further by including terms for those paved and pervious areas that are specifically designed not to drain to any point on the network or the river. This would be developed as follows:

Additional volume of runoff	= Paved areas draining to the network
	+ Pervious runoff draining to the network or river
	– Pervious area of the greenfield site.

This can be developed in the following form:

$$Vol_{xs} = RD.A.10\left[\frac{PIMP}{100}(\alpha 0.8) + \left(1 - \frac{PIMP}{100}\right)(\beta.PR_{RURAL}) - PR_{RURAL}\right]$$

where:

Vol_{XS}	is the extra runoff volume (m^3)
RD	is the rainfall depth usually defined by using duration T_p or 12 hours (mm)
PIMP	is the impermeable area as a percentage of the total area (0–100)
A	is the area of the site (ha)
SOIL	is the "soil" index from the WRAP map (values from 0.15 to 0.5)
α	is the proportion of paved area draining to the network or directly to the river (0–1)
β	is the proportion of pervious area draining to the network or directly to the river (0–1).

If one examines the equation PR_{RURAL}, however, and in particular the indices used in the term SPR, which uses constants that are virtually the same as the SPR values of SOIL, this equation can be simplified to:

$$Vol_{xs} = RD.A.10\left[\frac{PIMP}{100}(\alpha 0.8) + \left(1 - \frac{PIMP}{100}\right)(\beta.SOIL) - SOIL\right]$$

α and β are terms defining the proportion of paved and pervious areas draining to the network. Therefore, if all the paved area is assumed to drain to the network and all the pervious areas are landscaped not to enter the drainage system or river, this formula simplifies to:

$$Vol_{xs} = RD.A.10\left(0.8\frac{PIMP}{100} - SOIL\right)$$

But where all pervious areas are assumed to continue to drain to the river or network, the formula becomes:

$$Vol_{xs} = RD.A.10\left(0.8\frac{PIMP}{100} - \frac{PIMP}{100}.SOIL\right)$$

It should be noted that the FEH uses the index HOST for soil class, but fortunately the same concept of an SPR value exists for each HOST class. Thus this formula is equally applicable to using HOST or SOIL.

Figure 10.5 illustrates the additional runoff volume to be stored for developments with different soil types and development densities, assuming all pervious areas are specifically prevented from draining directly to the network or river and all paved areas are positively drained. This shows that even with SOIL type 1 careful use of landscaping results in very little storage being needed. It also emphasises the importance of the degree of development in terms of impermeable surface.

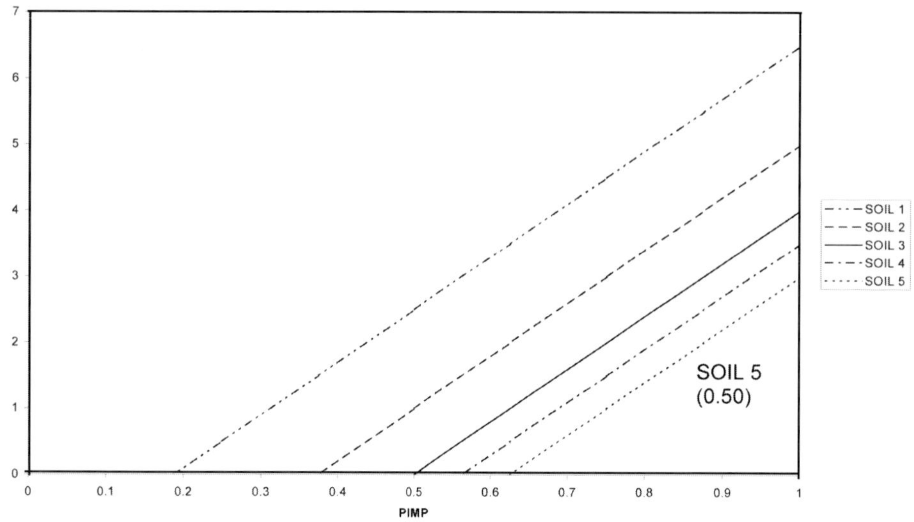

Figure 10.5 *Additional runoff caused by development with all pervious areas positively not drained*

Conversely, the following graph (Figure 10.6) shows that if all of the pervious area continues to drain directly to the river, storage volumes need to be much greater for developments with low volumes of PIMP. As the development tends towards being 100 per cent paved, the storage required will tend to the same volume.

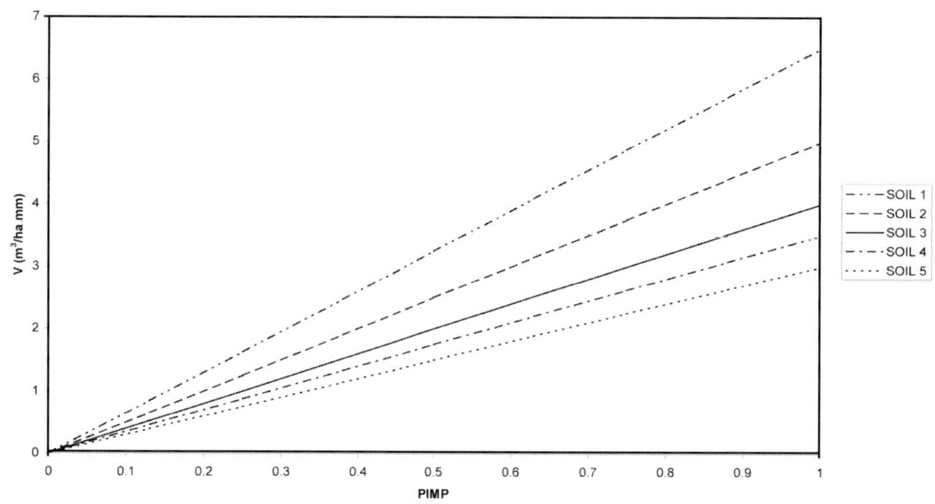

Figure 10.6 *Additional runoff caused by development with all pervious areas assumed to be positively drained*

This equation is conservative in not allowing for depression storage and making the assumption that all pervious areas continue to contribute as in the undisturbed state, but it does provide a conservative and rapid method of assessing the maximum additional volume of runoff generated by a development.

The current requirements and also the proposed procedure, which have been defined in this chapter, are illustrated using two examples in Appendix D. The examples do not use the alternative approach for estimating excess runoff volume, but use the equations from the Wallingford Procedure and FSSR 16 (IH, 1985).

10.5.3 The calculation of T_p for river flood protection analysis

The proposed methodology suggests that a rainfall duration of six hours should be used. In theory, a more correct method would be to use a time that is a function of T_p, which relates to the critical peak flow of the river at that point. However, it is felt that a pragmatic value that would be effective for all rivers would be easier to apply, although for the sake of completeness information on T_p is provided. This allows an alternative approach to be taken where it is thought to be appropriate. Note that this is not the same as trying to determine the critical duration storm for the catchment, which is around $2T_p$. T_p can be loosely translated as the time of concentration for urban drainage engineers.

From FEH:

$$T_p(o) = 4.270\ \mathrm{DPSBAR}^{-0.35}\ \mathrm{PROPWET}^{0.80}\ \mathrm{DPLBAR}^{0.54}\ (1 + \mathrm{URBEXT})^{-0.77}$$

Where:

DPSBAR	Mean drainage path slope
DPLBA	Mean drainage path length
PROPWET	Index of proportion of time that soils are saturated, based on estimates of soil moisture deficit
URBEXT	FEH index of fractional urban extent, judged from digital maps of land cover at 50 m intervals.

In general, $T_p(o)$ will range from around 4 hours for small catchments at the head of rivers through to 24 hours or more. Most developments will be on small watercourses and streams due to the nature of topography. On rivers such as the Thames, T_p can be much longer than 24 hours. As the storage volume is a direct function of rainfall depth, which increases with duration, this means that two sites with the same level of development on the same soil type would require very different river flood protection storage if their catchment sizes differ significantly. The logic of using a factor related to T_p cannot really be contested, as river flooding is associated with the duration of the rainfall. However, this would result in all developments adjacent to large rivers being far more expensive in terms of storm drainage provision than those on minor rivers. For this reason, and also for simplicity and pragmatism, it is suggested that a value of 6 hours be used. This is generally the range for the calculated value of T_p for small catchments and provides a reasonable basis for a realistic requirement for storage assessment to be effective for any catchment.

10.6 Effects of storage design on tank size

There are practical difficulties in meeting hydraulic criteria, as adopting authorities rarely accept orifice controls or pipe sizes with diameters less than 150 mm. Although there some vortex devices can reduce the flow through a throttle unit yet still provide a free bore of 150 mm, developments below a certain size will not be able to throttle the flow sufficiently to meet the stated criteria. Figure 10.7 illustrates this limitation based upon the hydraulic capacities of different pipe sizes. Hydraulic vortex units are often used, but these tend to approximate to equivalent pipe diameters of a smaller pipe, so equivalent curves can be developed.

Therefore, where long-term flood storage is required, the use of a throttle is unlikely to be practicable for many development sites, as the orifice would be too large to achieve throttle rates in the order of 1 or 2 l/s/ha. This indicates that the volumetric element of river flood protection is likely to be achieved by other methods, particularly infiltration, to ensure that part of the runoff from impermeable surfaces does not pass directly to the drainage system.

The criteria in Table 10.1 of a minimum discharge rate is based on the premise that the minimum orifice and pipe size needs to be 150 mm and self-cleansing flows should be at least 10 l/s. In addition, it is felt that 10 l/s is a practical lower limit to not warrant worrying about catchment impacts. It should be stressed that various SUDS techniques do achieve lower discharge rates and should be used where appropriate.

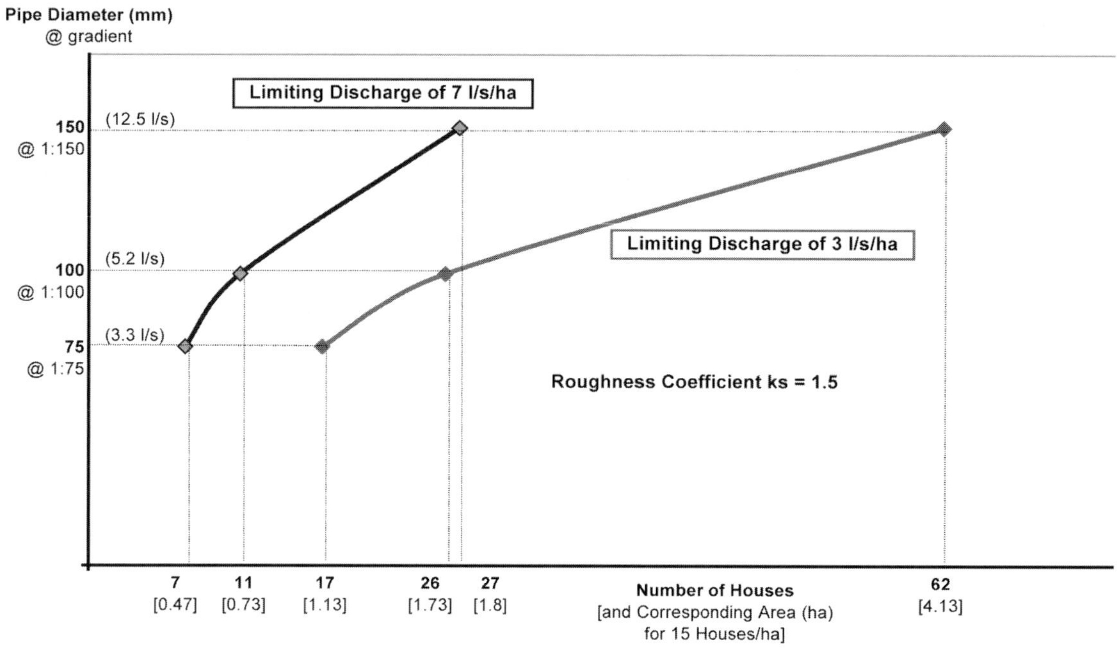

Figure 10.7 *Throttle size limitations related to development area*

Accuracy of tank sizing

Sizing of tanks is usually carried out assuming a fixed discharge limit. This discharge limit may be more than one value to meet various return period criteria. In practice this could result in a plethora of orifices at various levels in a tank, or variable weir profile controlling pond outflows. It is important to make sure the final design is practical as well as effective.

The head-discharge relationship of pipes (and other structures) requires accurate modelling. The difference between an assumed fixed discharge rate and a variable one can result in the tank size being increased by up to 30 per cent. It is therefore important to consider the depth/volume storage relationship and filling methods in order to optimise the tank/pond design. Designing offline storage has obvious advantages in this respect in minimising volume requirements, although these units have other limitations in terms of water quality benefits and operation and maintenance aspects. The head-discharge relationship is often complicated by the fact that the receiving water body into which a storage system might discharge has varying water levels depending on the state of the river. This introduces the issue of joint probability, which must be considered if the storage is adjacent to the watercourse.

10.7 Summary of methods for calculating storage requirements

The following statement provides the requirements of the current and proposed methodology for the design of stormwater storage.

Current methodology	
• Determine limiting discharge	• Calculate storage requirement
Proposed methodology	
• Determine range of limiting discharge rates • Determine runoff volumes for greenfield and development sites	• Calculate interception storage • Calculate attenuation of storage using greenfield runoff rates • Calculate "long-term" storage for additional runoff volume

10.7.1 Current methodology

The flow chart in Figure 10.8 summarises the approach currently used in calculating storage.

It should be noted that the mean annual flood or one-year event that is usually specified is usually a low rate of discharge, especially for sites with sandy soil characteristics, and this can result in very large storage volumes being calculated.

The flow chart presumes that calculations are carried out using FSR or MAFF methods. It does not refer to COPAS (Copas, 1957), the Rational-based methods or any of the other empirical approaches that some engineers still occasionally employ.

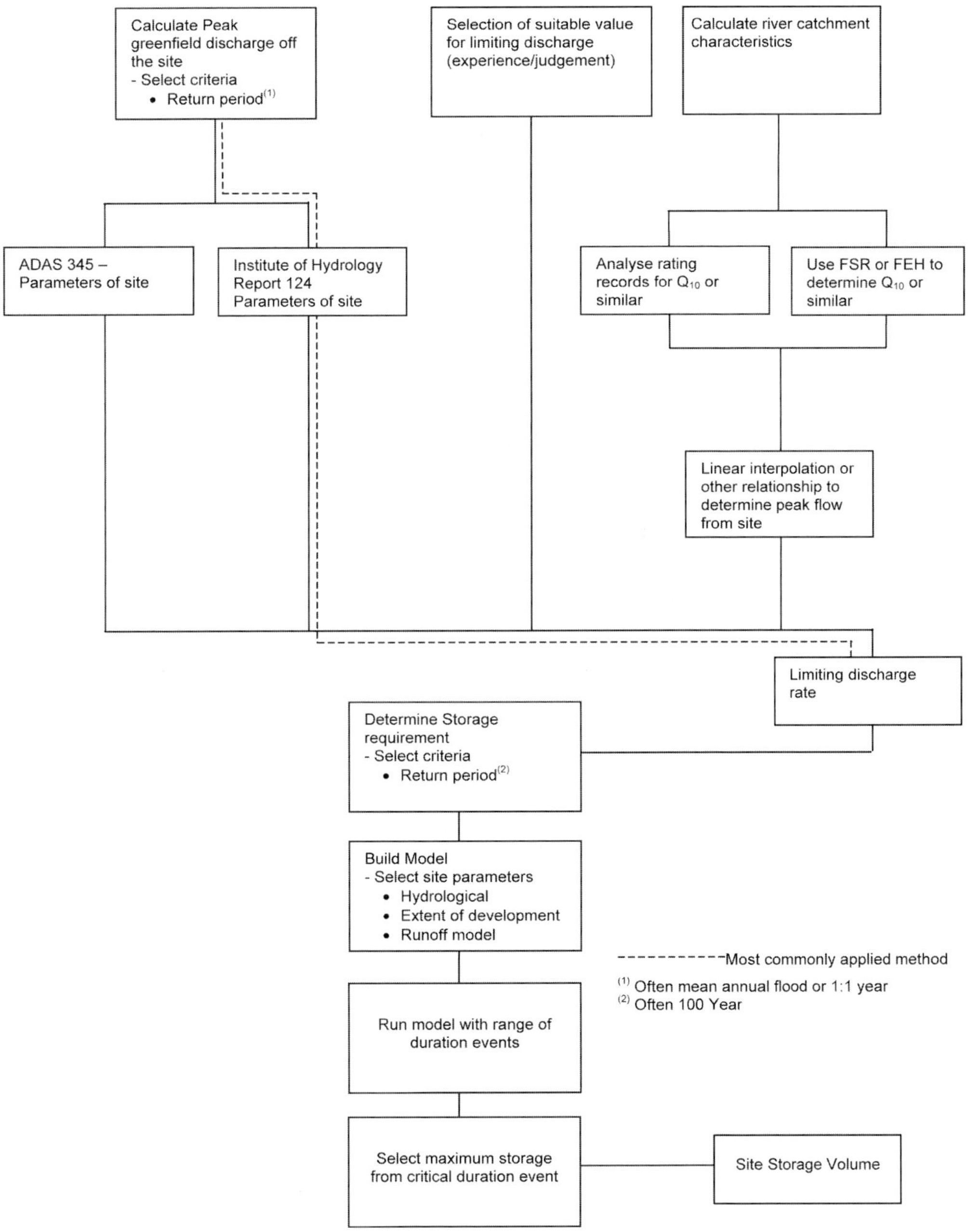

Figure 10.8 *Current methods for determining storage volume*

10.7.2 Proposed methodology

The following flow charts (Figures 10.9–10.14) illustrate the requirements for the proposed method when using a system of throttles and overflows. In practice, a mix of SUDS units will be used to achieve the same effects. Because there are up to five levels of storage, the process is more extended but uses the principles and approach of the current procedure.

Figure 10.9 illustrates the process that is followed to determine the five storage volumes rather than the one step used by the existing process.

It can be seen in Figure 10.9 that a check is carried out after the calculation of Stage 4 (river flood protection). Stage 4 is the determination of the long-term storage requirement. If the site system of storage proposes to mobilise this storage volume using a range of tanks or ponds, it is quite possible that the long-term storage will not be properly utilised by first designing for the "level of service" storage for the site. If this is the case it would be necessary to mobilise the long-term storage at an earlier stage, ie after the river regime storage is utilised and before the "level of service" storage comes into effect. This situation particularly occurs for SOIL types 1 and 2. Catastrophic flood impact is not shown, as this is unlikely to need analysis for most developments. Where catastrophic risk is deemed to exist, solutions are not necessarily linked to the storage systems derived for the first five criteria.

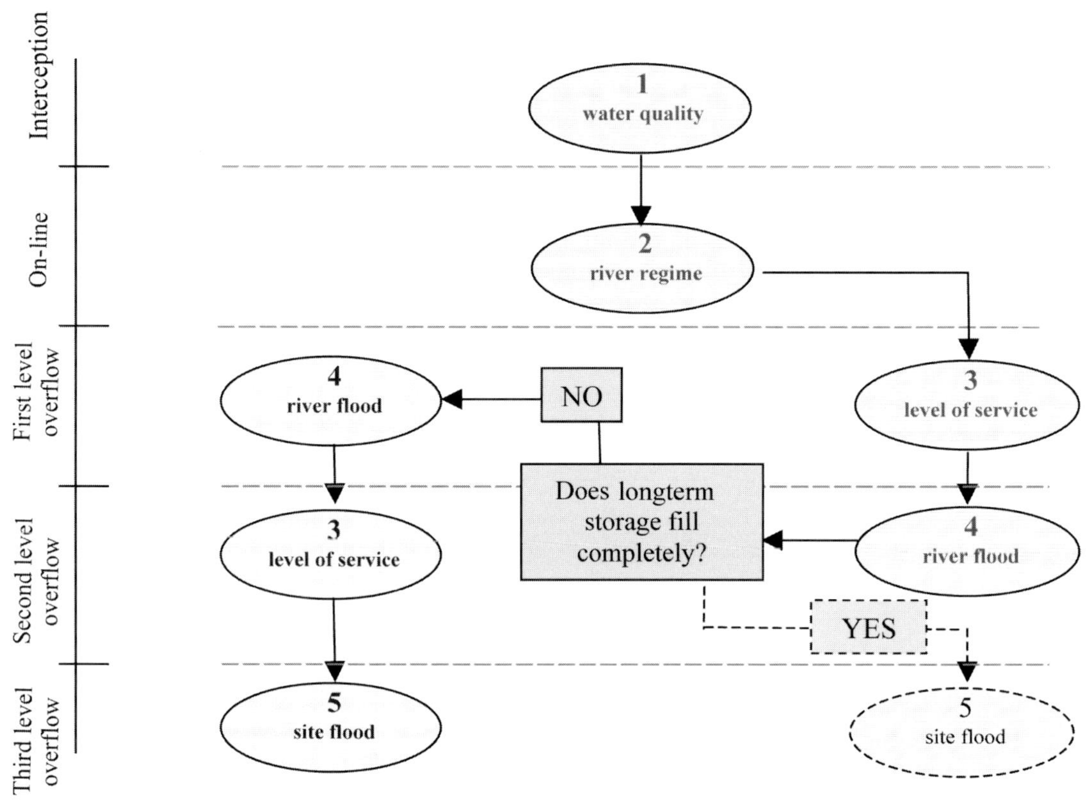

Figure 10.9 *Flow path of proposed methodology for storage volume design*

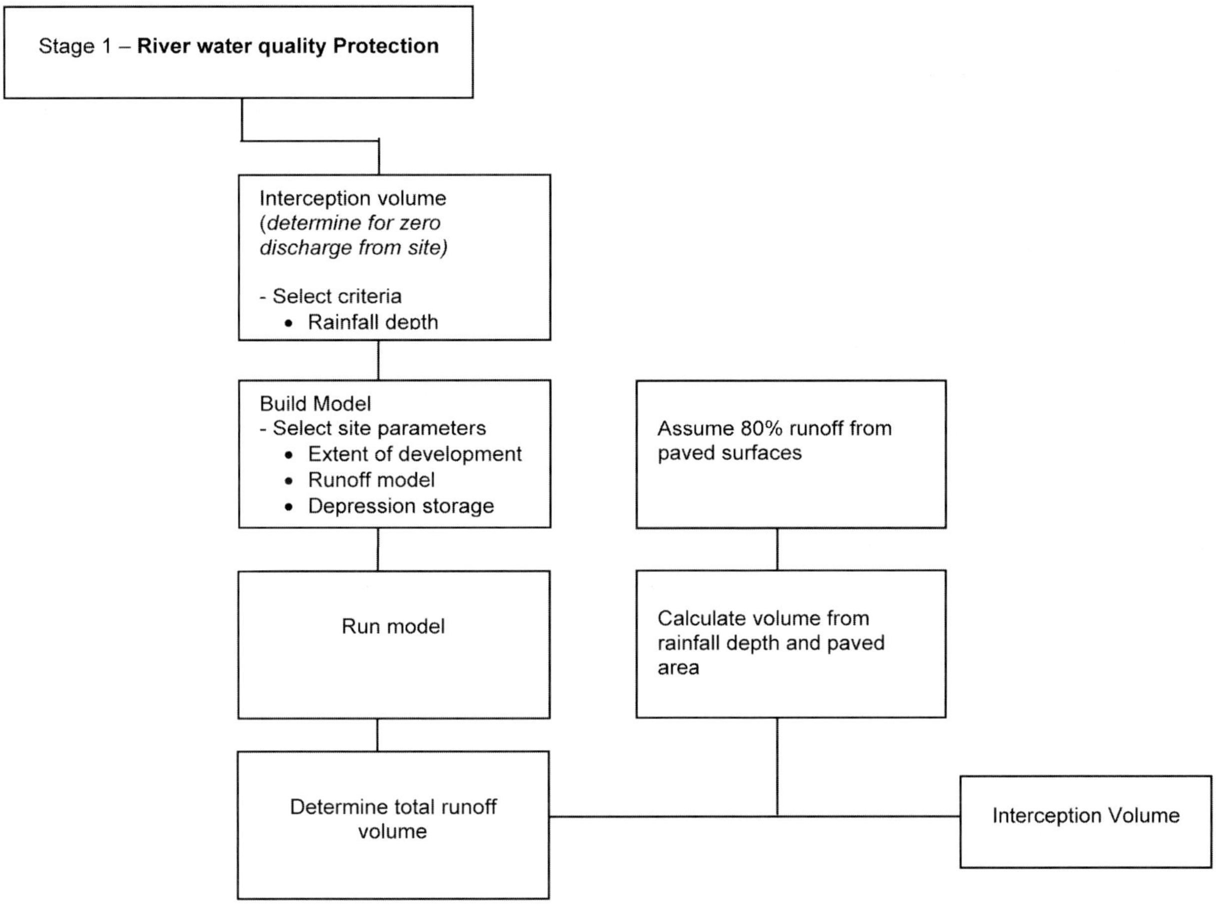

Figure 10.10 *Proposed storage design methodology – Stage 1, Water quality*

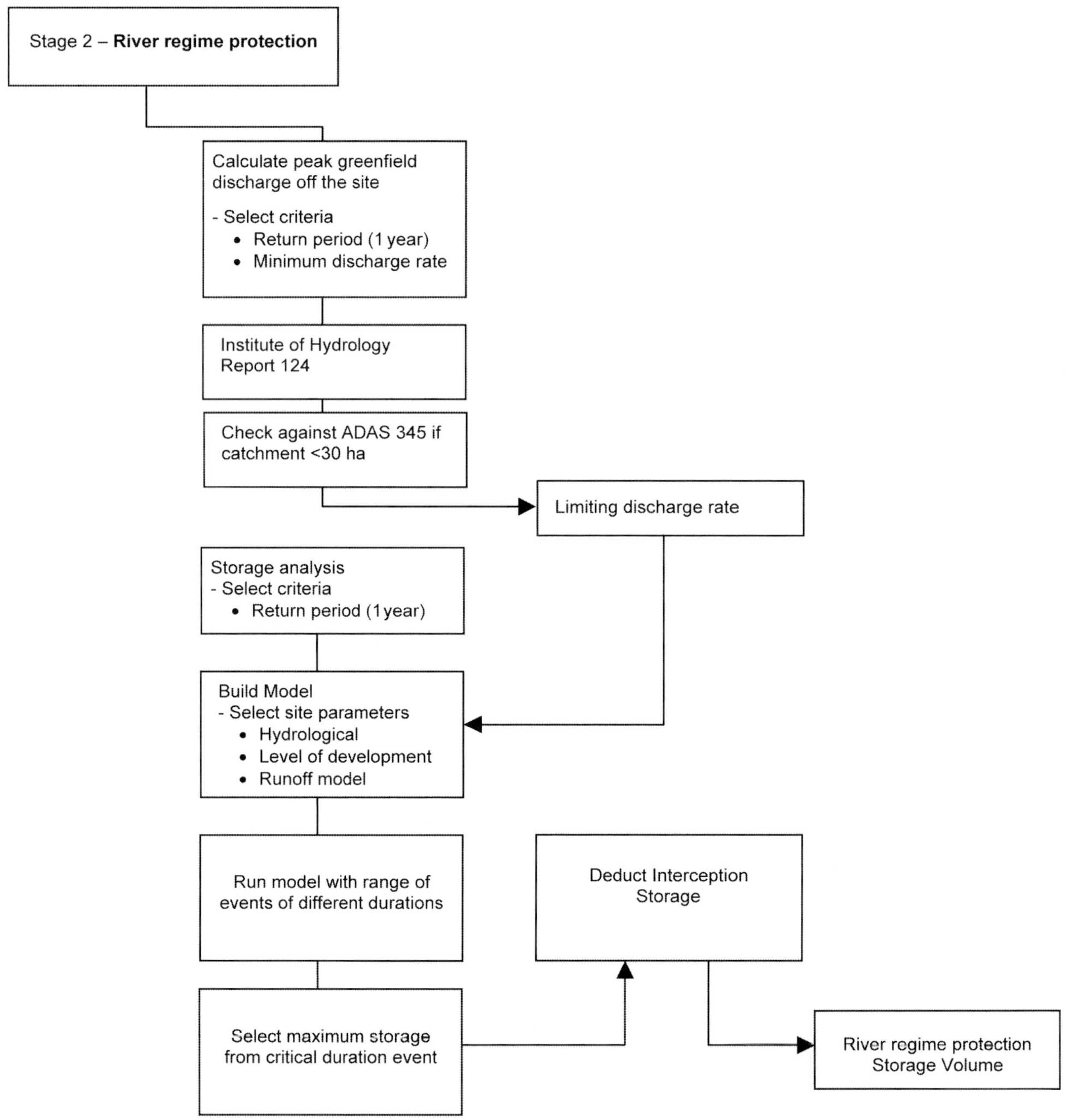

Figure 10.11 *Proposed storage design methodology – Stage 2, River regime protection*

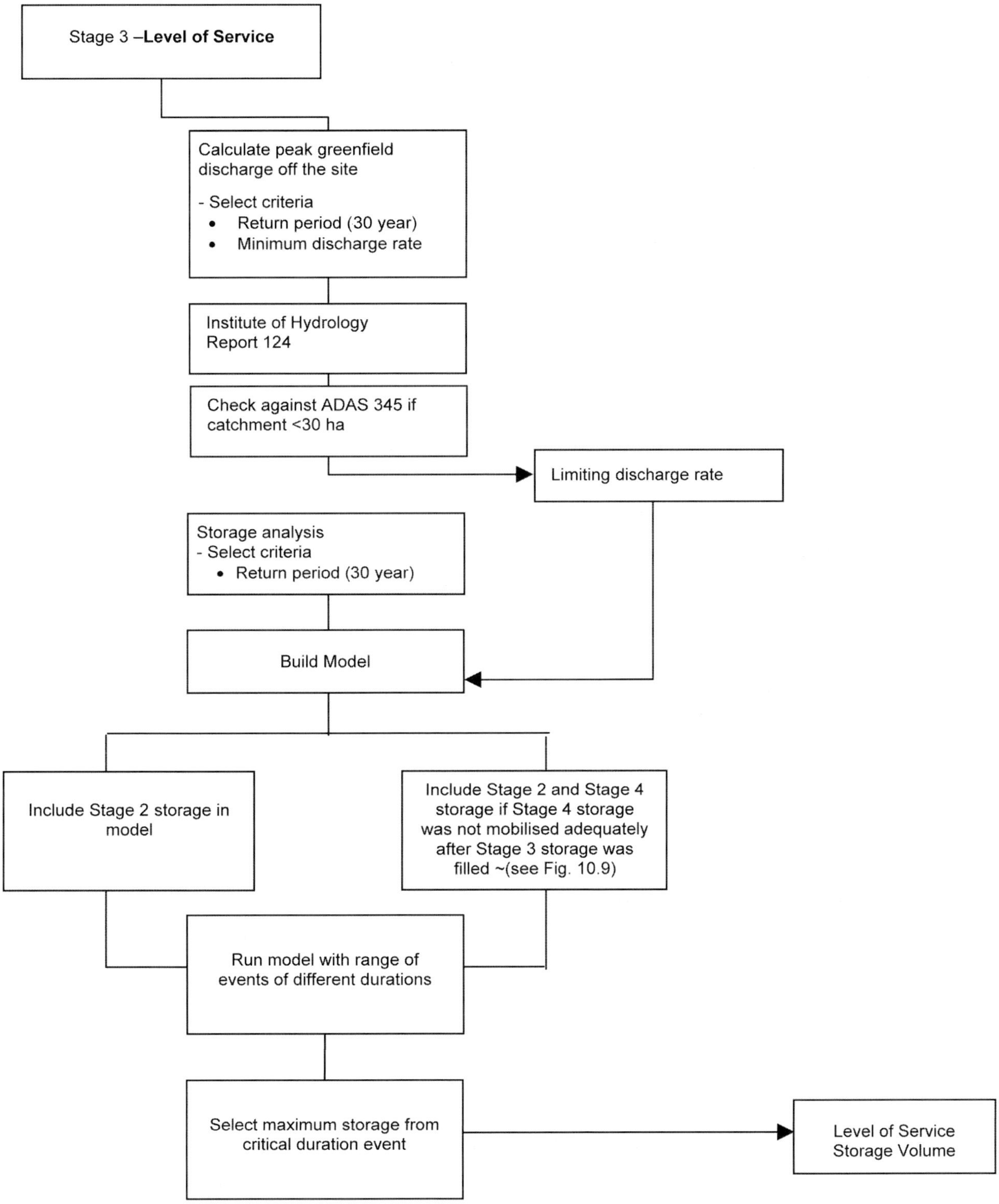

Figure 10.12 *Proposed storage design methodology – Stage 3, Level of service*

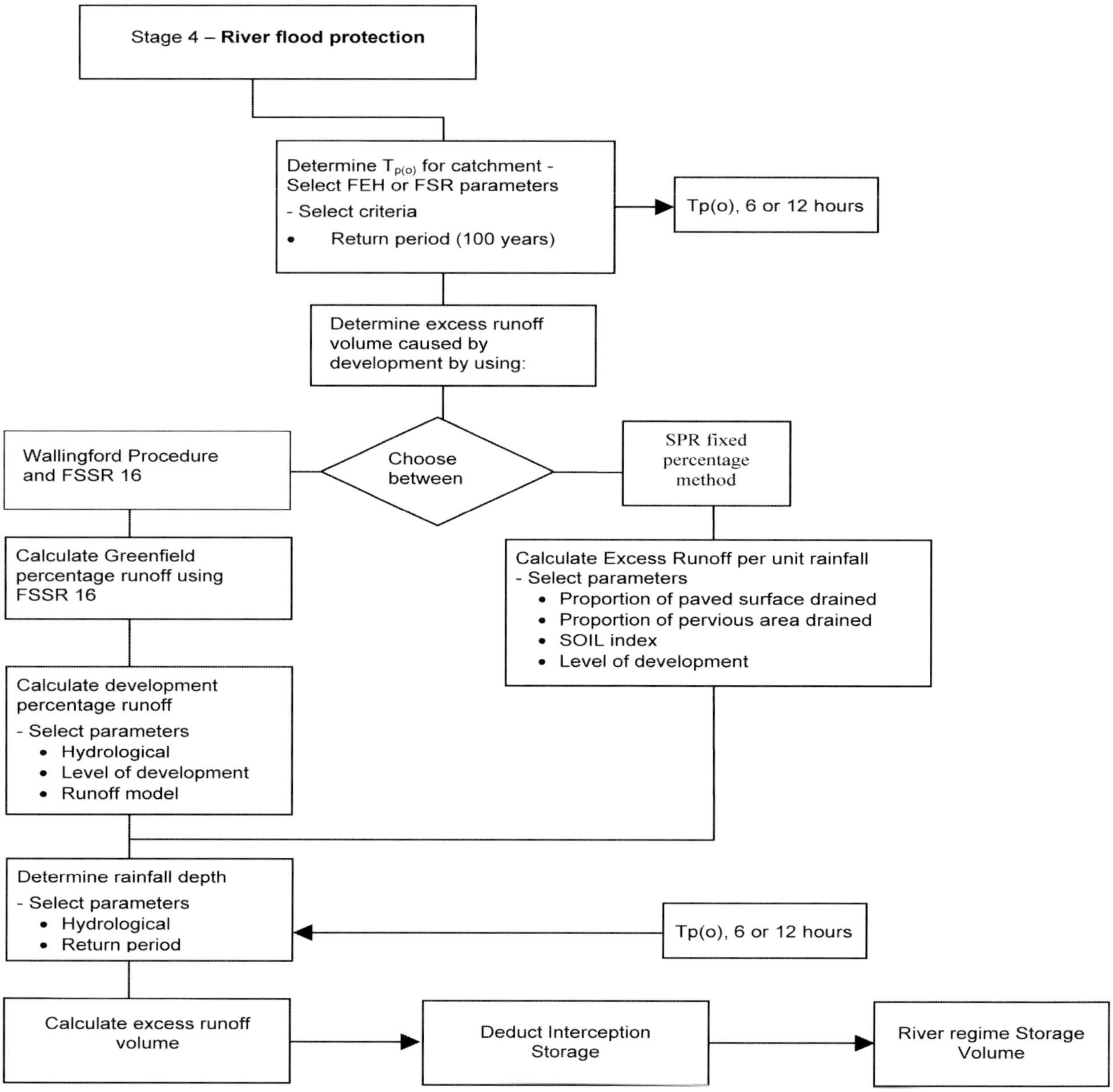

NOTE: *This flow chart is different from all the other flow charts as the calculation is based on volumes of runoff and rates of runoff.*

Figure 10.13 *Proposed storage design methodology – Stage 4, River flood protection*

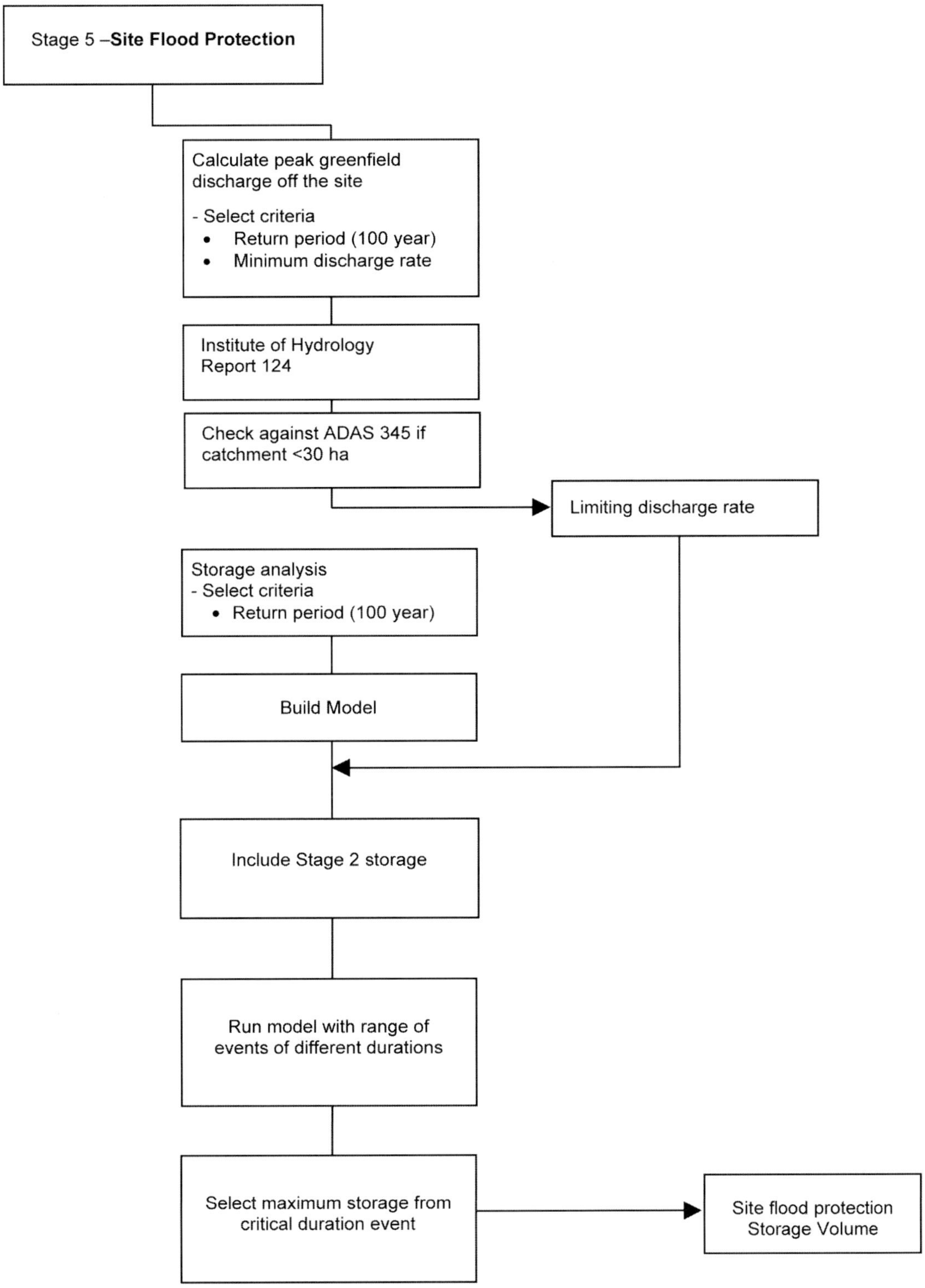

NB(1) This flow chart is similar to the level of service flow chart (Stage 3) in that peak greenfield runoff rates are calculated and a model is built which includes the storage volumes evaluated for Stages 1 to 4.

NB(2) This storage is unlikely to be built as a formal storage pond or tank, but takes the form of car parks or park areas.

Figure 10.14 *Proposed storage design methodology – Stage 5, Site flood protection*

APPENDICES

A ORGANISATIONS AND REGULATORY STRUCTURES

The number of authorities involved in drainage and their various roles and responsibilities often causes confusion. References to "the Water Board" are still regularly encountered in general conversation, which illustrates both the speed of change that has taken place over the last few years and the gap in the public's knowledge in this area. The following chapter briefly describes the roles and responsibilities (that are relevant to site drainage) of the various authorities involved. These are:

- local authority
- sewerage undertaker
- Environment Agency/Scottish Environmental Protection Agency/internal drainage board
- Office of the Deputy Prime Minister (DETR)/DoE (NI)/Welsh Office
- Office of Water Services (OFWAT)
- Highways Agency
- owners (British Waterways Board, Network Rail etc).

Reference documents and contact details are also listed to allow more information to be obtained.

A1 The local authority

Local authorities are the principal points of contact for obtaining planning approval. All the other bodies mentioned here have drainage powers and influence the planning process.

A1.1 District and borough councils

The district council is the local planning authority and building control authority (Building Act 1989). It co-ordinates the responses of other agencies with regard to the drainage aspects of any planning proposal.

District councils have wide-ranging powers under the Public Health Act 1936 and succeeding legislation, the Land Drainage Act 1991 and the Local Government Act 1972 with respect to land drainage on non-main rivers (ordinary watercourses). These relate particularly to flood prevention and maintenance of watercourses. In practice, the local authorities look to the Environment Agency for guidance and advice on any regulatory control that may be possible, particularly under the T&CPA conditions that they may impose.

A1.2 County councils

The county council is not usually involved in the standard planning process except in regard to mineral and waste applications and school sites. County councils have two drainage roles: as a highway authority and as a land drainage authority. The latter role is defined by the 1991 Land Drainage Act. This empowers the council to execute land drainage schemes and to ensure that riparian owners maintain watercourses. Trunk road drainage is not included, as this is the responsibility of the Highways Agency and the Welsh Office.

Under the Highways Act 1980, the county council is responsible for preventing roads from flooding and has powers to:

- adopt a highway drain
- prevent water flowing both on to a highway and from the highway
- require developers to obtain consent for any works or use of a watercourse for highway drainage.

The county council may have an interest if it is proposed to divert a significant flow to a non-main river watercourse. It can also become involved in place of the local authority by default.

In the context of general planning, the county council defines the structure plan. This document is linked to the local development plan, which defines proposed future land use.

A1.3 Unitary authorities and metropolitan borough councils

Unitary authorities and London boroughs possess an amalgam of the powers of both the district and county councils.

A1.4 Parish councils

The parish council has no administrative role with regard to watercourses or land drainage. However, it is likely to have views and make representations to the local authority on issues such as ponds and swales.

A1.5 Regional councils (Scotland)

Prior to the creation of the water authorities in Scotland, the regional authorities had a two-tier organisational structure, within which the wider regional councils had responsibility for water and sewerage. Today there are 32 unitary authorities, which are responsible for all planning issues including the consenting of surface water discharges to rivers. SEPA is responsible for the water quality aspects of discharges, and there is considerable co-operation on drainage issues between these two organisations.

A2 The sewerage undertaker

A2.1 Water service companies in England and Wales

The Water Act 1989 split the water authorities in England and Wales into 10 water service companies, generally referred to as sewerage undertakers, and the National Rivers Authority (NRA), now the Environment Agency. The sewerage undertakers are responsible for the provision of water and drainage services. Their responsibility is for the foul and surface water pipework, which are defined as public sewers and are shown on maps held by the sewerage undertakers and local authorities.

Although the sewerage undertaker has a responsibility to provide a drainage service, it also assesses planning proposals for their impact on the system downstream and can object to a proposal. The local authority usually takes note of the objection, which it will discuss with the developer. The develop may agree either to provide a requisitioned sewer or, through appropriate on-site measures, to limit flows into the sewer system.

Sewers for adoption (WRc, 2001) specifies generally agreed standards of the drainage required to ensure that the sewerage undertaker can take ownership of the sewers.

A2.2 Scottish Water

The Scottish Water Authority is the sole organisation responsible for the supply and disposal of water in Scotland. It was created in 2002 from the amalgamation of three organisations that had been set up by the Local Government etc. (Scotland) Act 1994 to supply water and sewerage services. It has a similar remit to the sewerage undertakers in England and Wales, although it remains a government organisation and has not been privatised.

A3 The Environment Agency

The Environment Agency was created by the Environment Act 1995, which amalgamated the roles of the NRA, Her Majesty's Inspectorate of Pollution (HMIP) and waste regulation authorities. The NRA had itself been formed by the Water Act of 1989, which split the water authorities into water service companies and the NRA.

One of the Agency's principal responsibilities – dating back to the Land Drainage Act of 1930 – is "to exercise a general supervision over all matters relating to flood defence".

The Environment Agency has "permissive powers" for all those watercourses that DEFRA designates "main rivers" (under the Water Resources Act 1991). Permissive powers enable a statutory authority to authorise the Environment Agency to carry out necessary maintenance and improvement works. Although the Environment Agency is excluded from this role for non-main rivers, Circular 30/92 (DoE, 1992) directs planning authorities to consult the Agency on all matters related to runoff into watercourses and proposed development liable to flood. In addition, local and highway authorities need to consult the Environment Agency before exercising their responsibilities under their general drainage powers.

The Town and Country Planning Act 1990 authorises the imposition of conditions on a planning permission with regard to drainage proposals. The Environment Agency seeks to regulate proposed works to watercourses under the provisions of the Water Resources Act 1991, which relates to main rivers, and the Land Drainage Act 1991, which relates to "ordinary" watercourses (ie, those other than "main rivers").

Discharge consents for outfalls are issued on a discretionary basis for surface water discharges to "controlled waters" under the Water Resources Act 1991. Under the same Act, land drainage consents are also issued for structures, including outfalls, which could affect flows in main rivers. Consents for structures that could affect flows on ordinary rivers come under the provisions of the Land Drainage Act 1991. The control of discharge quantities is generally advisory, and the Agency seeks to work in partnership with local planning authorities to regulate stormwater drainage through planning conditions and/or agreements.

The Environment Agency also has powers to carry out works for the purpose of coastal defence both above and below the low water mark. These powers do not extend to coastal protection against erosion, for which the coastal local authorities are responsible.

Planning Policy Guidance Note 25 *Development and flood risk* (DTLR, 2001) particularly emphasises the use of SUDS systems. It replaces Circular 30/92 (DoE, 1992) and forms a major element of Environment Agency policy on development drainage. The document uses the concept of the "precautionary principle" and draws attention to the potential impact of climate change. It lays on the developer the responsibility for preventing flooding, both on site and downstream, caused by construction.

Figure A1 *Environment Agency areas and regions*

A4 SEPA

The Scottish Environmental Protection Agency was created by the Environment Act 1995 and came into being in 1996. SEPA was created from the seven river purification boards and three island councils and has responsibility for water quality, waste regulation and air pollution in Scotland.

A5 Internal drainage boards

In England and Wales there are 250 internal drainage boards. They deal with ordinary watercourses and exist to provide protection to low-lying areas subject to flooding as well as to maintain adequate drainage. The water levels in the drainage systems are commonly controlled by pumping. Boards are run by elected members who are local residents or have agricultural property within the IDB area.

The linkage between the IDBs and the Environment Agency is defined in the Land Drainage Act 1991. The Environment Agency can give advice and direction and it must be involved when proposals involve more than one IDB. It can, under some circumstances, take over the running of an IDB.

A board can levy on a developer contributions towards upgrading or modifying the drainage system or for the additional cost of running the network.

Figure A2 *IDB drainage areas*

A6 ODPM, the Welsh Office and DoE (NI)

In Northern Ireland, all responsibility for water and sewerage lies with the DoE (NI). It is subdivided into four regions: North, South, East and West. The regulator deals with all planning issues relating to the discharge of stormwater.

ODPM (formerly DETR) and the Welsh Office have strategic planning and administration responsibilities for water affairs in England, Scotland and Wales. They issue circulars and guidelines that provide policy guidance.

Where the responsibility has not been devolved down to local authorities, the Welsh Office has a specific interest in the drainage of trunk roads.

Developers who have been refused planning permission by the local authority can appeal to the ODPM.

A7 OFWAT

The Office of Water Services (OFWAT) was created by the Water Act 1989 to exercise control over water companies. Its role is to safeguard customers by ensuring that water charges are fair and that the level of service provided is in accordance with the charges that are levied. It may be consulted and appealed to in the case of a dispute.

A8 Highways Agency

The Highways Agency is responsible for motorways and those trunk roads in England that are not the responsibility of local authorities. They have rights of drainage without requiring consent to discharge surface water runoff.

A9 Owners

The **British Waterways Board** is a nationalised industry body sponsored by DEFRA and the Scottish Executive. It was formed in 1963 from the British Transport Commission. Its roles are stipulated in the 1962 and 1968 Transport Acts.

British Waterways is not a statutory drainage authority and so is not obliged to accept surface water discharge. Commercial negotiated licences with both a capital charge and annual fees are usually stipulated.

Network Rail (successor to Railtrack) owns the land over which railway track (permanent way) is laid. Permission to pass drainage under the permanent way must be obtained. This can be very expensive.

The **National Trust** is a private organisation, which has interests in waterways and watercourses, and its agreement must be obtained for surface water discharges to these water bodies.

Approval should be obtained from the **Ministry of Defence** (a government department) for discharge to any waterway for which the MoD is responsible.

B VACUUM SEWERAGE

B1 Introduction

A vacuum sewerage system is a drainage system that operates under negative (sub-atmospheric) pressure to evacuate wastewater flows from a property or group of properties. The system may consist of one or more vacuum pumps, a central vacuum reservoir, pipework and interface valves.

It can be thought of as a system operating in reverse to a standard drainage system. Whereas a standard system drains "down", a vacuum system drains "up" from the discharge point. A vacuum sewerage system failing is inherently more problematic than failure of a standard system, as the flows flood back to individual properties at low points in the system.

B2 Uses of vacuum sewerage systems

These systems are generally used where the requirement for minimum pipeline gradients would cause difficulties.

They are used in flat areas with low-density housing where wastewater volumes are low and a traditional drainage system does not exist. The use of a vacuum, instead of gravity, as the driving force for transport of wastewater, allows small-bore pipes to be laid across long distances at shallow depths beneath the surface. This latter aspect also makes the system attractive for use in areas where it might be prohibitively expensive or technically difficult such as digging deep trenches or tunnelling, confined areas, large industrial sites with high-density infrastructure and solid rock.

Buildings such as hospitals or sports arenas also often use the systems, because their toilet and washing facilities may be scattered widely across the site, or between floors in extensive buildings. Vacuum systems minimise the space needed for pipes and provide greater flexibility for designers, as pipes do not have to be laid to self-cleansing gradients.

Vacuum toilets are often installed in buildings located where there is a need to control and conserve the amount of wastewater produced in each flushing cycle, as systems can transport wastewater with a high solid-to-liquid ratio.

B3 Advantages of vacuum sewerage systems

Vacuum sewerage systems perform similar functions to common gravity pipe systems and pumped pressure pipe systems, but have important differences.

1. They can be used in permafrost conditions that preclude the use of an over-ground system.
2. They are suitable for transporting hazardous or contaminated fluids, such as wastewaters or sewage, because of the need for sealed pipe systems and negative pressures.
3. There is a high level of flexibility in possible design options (eg wastewater can be lifted vertically up to 7 m from the point of collection). The possibility of lifts and lateral transportation without the loss of height gives freedom to plan the usage of the building space more efficiently, as appliances producing wastewater can be placed anywhere and are not tied to the vicinity of a gravity drain.
4. The systems essentially eliminate the need for under-floor piping within building.
5. As vacuum systems are often used in conjunction with water-saving appliances; reduced wastewater volumes allow for smaller pipe diameters, potentially reducing installation, maintenance and future modification costs.
6. Wastewater is transported at around 4 m/s, which limits the risk of pipe obstructions and reduces the possibility of bacterial growth.
7. The option of multiple collection tanks can be used to different types of wastewater, potentially minimising treatment costs and optimising the reuse of non-contaminated wastewater streams.

B4 Disadvantages of vacuum sewerage systems

Disadvantages of vacuum systems are as follows.

1 Energy demands are high, as they lack the natural force of gravity to aid transport of wastewaters.

2 The consequence of failure of the system is generally more severe. Failure of the vacuum system can result in flooding of houses and other collection low points on the network.

3 The systems are highly sensitive to overload from excess flows. If the system is not completely sealed – for example, if collars at pipe joints in the gravity sections of the collection systems are left off during construction, or if house roofs or other runoff areas have been connected to the system illegally – then rainfall or groundwater can drown out the system.

4 Vacuum systems tend to be more complex than traditional drainage systems, so higher levels of expertise are required not only for initial construction but, more importantly, for operation and maintenance. Air bleed times, for example, depend on the location of the valve in the system; and the failure of an interface valve is likely to lead to the whole system going down. Targeted telemetry installations and operational experience are vital in maintaining a high level of service.

5 Odour can be a problem. As the vacuum draws air through the system, it aerates the sewage/industrial wastewater and releases odours. Care should be taken in locating the vent at the vacuum pump, and air scrubbing may have to be considered in confined areas.

B5 The principles of the vacuum system

The system relies upon normal atmospheric pressure (air pressure) to force collected drainage fluids though pipework towards a collection vessel. This fluid is then pumped to a normal gravity main sewer or other specified destination.

To create a pressure differential so that atmospheric pressure at the inlets can be utilised as the driving force, negative pressure – a vacuum – has to be generated at the "outfall". This vacuum is created at the end of the system by means of vacuum pumps or eductors.

The drainage fluids collected in the various collection sumps are then "sucked" into the pipeline. However, to enable fluid to collect in a sump in the first place, the vacuum lines must be closed until the sumps have filled by gravity. Each sump has a vacuum interface valve connecting it to the vacuum piping system. This valve remains closed until the sump is full. When the valve opens, by means of a trigger based on the sump hydrostatic pressure, the pressure differential between the fluid in the sump (at "normal" atmospheric pressure) and the pipe, which is at a negative pressure (vacuum) of between -0.55 and -0.7 bar, causes the sump to be emptied. A typical layout is shown in Figure B1.

B6 Vacuum sewer profiles

Vacuum sewers are generally laid in a saw-tooth profile, as shown in Figure B2. The minimum drop between lifts is usually approximately 0.2 per cent of the fall distance. Where they are installed at 0.2 per cent fall in flat land, the trench depth should increase by 300 mm every 150 m. "Profile changes" are suggested at 150 m intervals to bring the sewer invert back to the starting level. This keeps the trench depth to a minimum. Where the natural ground profile has a fall in excess of 0.2 per cent in the flow direction, the vacuum sewer profiles follow those of the ground. The transport of the sewage is shown in Figure B3.

B7 The vacuum transport process

With the saw-tooth profile, and as long as no valves are operating, little sewage transport takes place. All sewage remaining in the sewers will lie in low spots, but without sealing the bore of the pipe, thus ensuring that little vacuum loss is experienced within the length of the system.

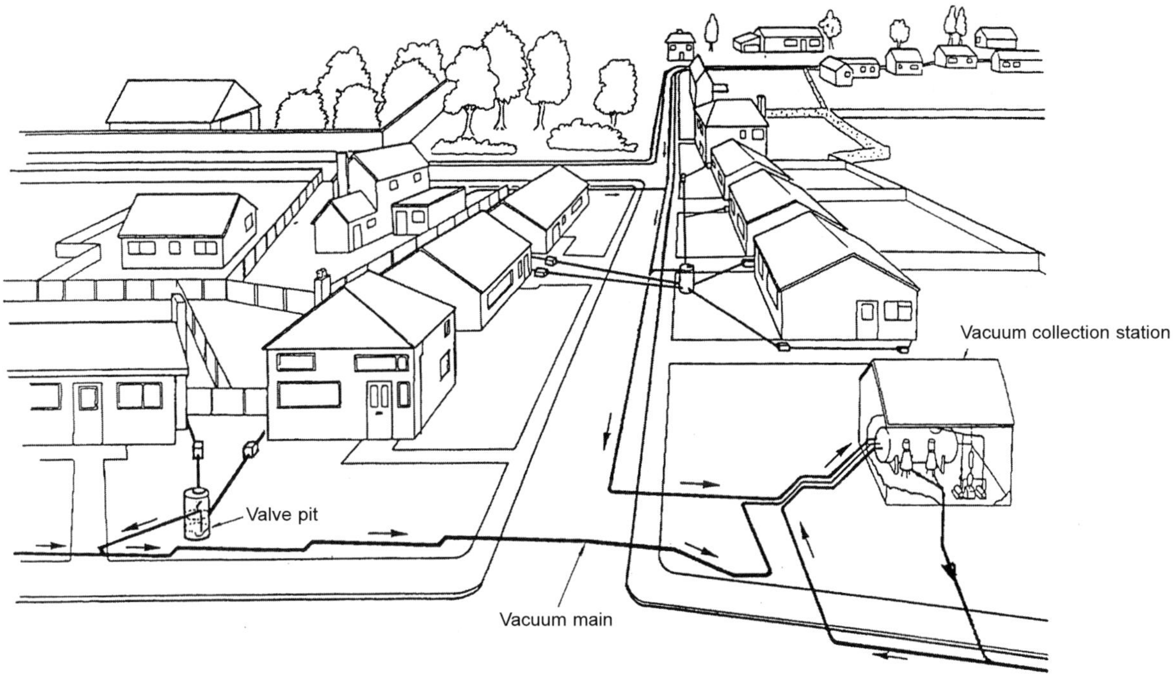

Figure B1 *A typical vacuum sewerage layout*

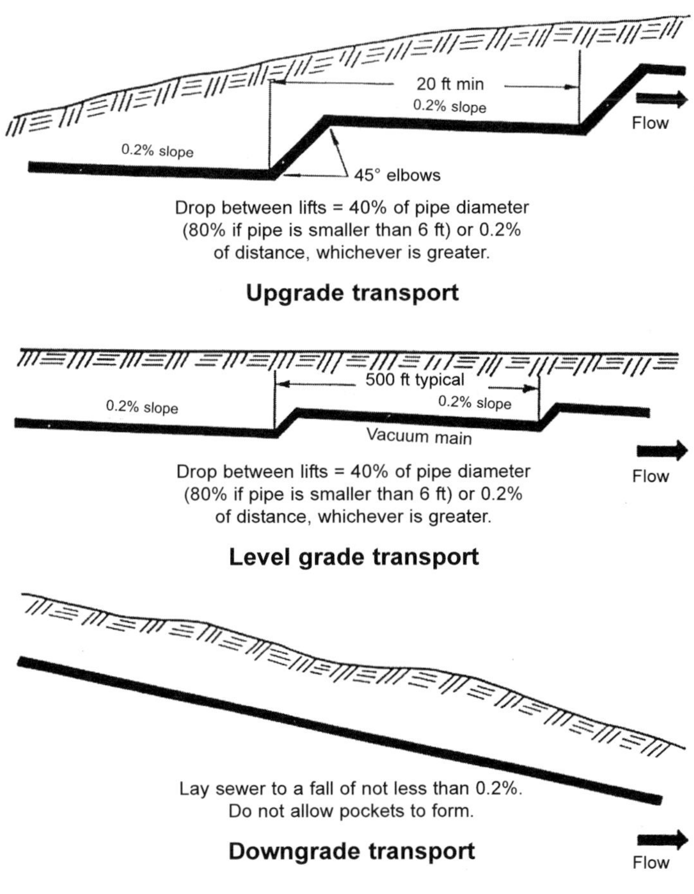

Figure B2 *Vacuum sewer profiles*

When the volume of sewage in a collection sump reaches around 40 l, its associated valve cycles. The differential pressure between the vacuum sewer and atmospheric pressure forces the 40 l of sewage into the vacuum main. While accelerating, the sewage is rapidly transformed into foam and soon comes to occupy only part of the sewer pipe cross-section so that the momentum transfer from air to water takes place largely through the action of shear stresses. The magnitude of the propulsive forces start to decline noticeably when the valves close but remain important as the admitted air continues to expand. Eventually friction and gravity bring the sewage to rest below several lifts.

Sewage admitted to a sewer through a vacuum valve initially moves in two directions. Some 80 per cent flows towards the collection station with the remainder flowing in the opposite direction. When the reverse, or backsurge, slows, gravity takes over and all the flow is towards the collection station (see Figure B2). The no-flow and backsurge processes are illustrated in Figure B3.

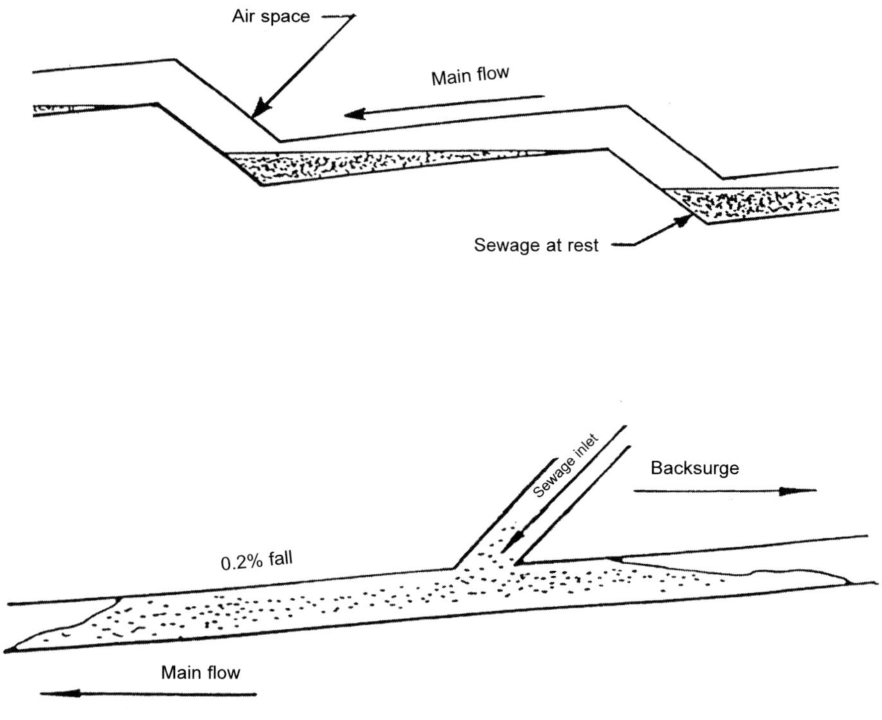

Figure B3 *Vacuum sewerage processes*

B8 Standards

BS EN 12109:1999 is the British and European Standard that deals with vacuum drainage systems inside buildings. The Standard specifies system performance requirements and the principal requirements for design and installation, with related verification and test methods, for vacuum drainage systems inside buildings transporting wastewater from dwellings and commercial properties, but excluding stormwater and rainwater. Annex D of the Standard provides guidance to performance requirements, design, verification and quality assurance.

B9 Principal design factors

The general performance of a vacuum drainage system is governed by its principal design factors:

- safety and health
- availability
- reliability
- maintainability

- noise and odour control
- energy economy
- fire resistance.

Each of these factors is dealt with in turn in the British (European) Standard detailed in Section 5 of BS EN 12109.

It is the designer's responsibility to ensure that discharges from the vacuum drainage system will not adversely affect the receiving drainage system, and to take any known future additions or modifications to the system into consideration to avoid future operating problems.

The following extracts from the British Standard cover the main design requirements for such systems.

1 The system must be designed so that the probable ***static loss*** at all times is less than the available vacuum differential, ie the difference between the system vacuum and the initial vacuum needed to operate interface units.

2 Unless specified otherwise, the expected ***wastewater flow rate*** shall be calculated using the probability calculation methods for gravity drainage systems.

3 The system shall be designed so that the ***dynamic losses*** of the line, at the design flow rate, calculated from the vacuum station to the end of the line, are less than the available vacuum differential.

4 The design of the system shall take the transient nature of the flow into account. Therefore, pipe diameters and piping profiles based on calculations of static and dynamic losses may have to be corrected in order to achieve shorter ***vacuum recovery*** times.

5 The ***system controls*** shall be designed, as a minimum, to maintain the system vacuum within the prescribed range and protect the equipment from flooding or running dry.

6 The ***status monitoring system*** shall, as a minimum, be designed to detect and indicate abnormal liquid and vacuum levels, abnormal vacuum generator running time and major equipment failures.

B10 Design criteria

The design of a system has to consider two principal features: the headloss of vacuum regeneration and the headloss of liquid flow. Design companies use various rules of thumb, but the intricacies of design are still a guarded secret.

The following points indicate some of the rules and system limitations when designing vacuum sewage systems.

<u>Limit on length of main</u>
The length of a vacuum main is a complex relationship between vacuum recovery time and the headloss related to the flow of liquid within it. Guideline maximum headloss figures provided range from 0.25 m to 0.5 m per 100 m for the dynamic state. Thus for a lift of 4 m (traditionally regarded as a limit in design five or 10 years ago), this suggests a maximum length of 1600 m. It should be stressed that systems up to 5 km and longer have been designed and operated satisfactorily, however.

<u>Pipe size</u>
A rough guide for pipe sizing of the vacuum main is 2 l/s for 75 mm, 4 l/s for 100 mm and 8 l/s for 150 mm for maximum peak flows. This is based upon dry-weather flow rates factored by 4, though figures of 4 to 6 are used. It should be noted that over-sizing of pipes does not cause significant problems in operating the system and provides significant improvements in vacuum recovery capability.

<u>Vacuum recovery</u>
Vacuum recovery is a function of pipe length, pipe size, vacuum pump size and liquid in the pipe. To ensure adequate vacuum recovery, 90 mm pipes are usually limited to a length of 100 m. Where water traps are formed (heavy flows or settlement of the main), the vacuum recovery can be significantly affected.

Limit on total lift
The maximum lift has traditionally been stated as being 3.5–4.0 m. In the past 10 years this has changed considerably with the use of timer air admittance valves. Figures of 10 m or more have been stated. Air admittance valves are used to assist in providing pressure differentials. Their use has to be considered in the pump sizing, but does not significantly affect it.

Static head is based on the total of all the lifts less the internal pipe diameters at each lift.

Collection sumps
Sumps are the points to which a number of domestic houses are drained in which an interface valve is located. The number of houses can vary between one and seven, although four is regarded as a norm. They are designed to collect about 40 litres, which is then injected into the vacuum sewer.

They are designed with a top water level based on several hours of DWF below the lowest incoming invert.

Design flow
The design of pipe size, vacuum pump sizes and gravity pumps are all based on factors of the dry weather flow. For normal domestic properties this is taken to be 600 l/property/day.

Air/liquid ratio
The air/liquid ratio is normally set at around 5:1. Achieving this is an art, as the level of vacuum at any interface valve ranges widely depending on its location on the main and the rate of vacuum recovery due to the current flow load.

The air intake is controlled by a screw acting as a needle valve to allow air to bleed into the control valve so that the interface valve stays open for between 2 s and 7 s.

Vacuum pump size
Vacuum pump sizes are based on six times dry weather flow. There are usually two pumps of both vacuum and gravity units. There are two types of vacuum pump: liquid-based and mechanical vane.

The liquid-based pump is cheaper, but often develops hard water (scale) deposits on the impeller. Use of cooling mechanisms only slows this process. The mechanical vane is more reliable and durable, but more expensive. However, if liquid gets into a mechanical vane pump, the failure is catastrophic, whereas the consequences are less serious for a liquid-based pump.

Gravity pump size
Gravity pump sizes are based on a flow rate of six to eight times dry weather flow. A guide of 10 per cent greater than the vacuum pump has also been used.

Vacuum tank size
The tank size is based on the volume on 15 minutes of average dry-weather flow. This is regarded as a minimum.

Location of collection sumps
Collection sump connections to the main should not be too close (within 2 m) to the downstream step (to minimise risk of reverse flow).

Junctions
Normally, the joining of tow pipes of the same size requires the pipe on the station side to be the next size up. Junctions should be Y-shaped and not T-shaped. Incoming pipe connections should drop into the main from above.

C ROOF DRAINAGE DESIGN EXAMPLE

C1 Worked example

A worked example is presented to illustrate the use of BS EN 12056-3:2000, in particular the application of the UK National Annexes to the European Standards for the design of a conventional roof drainage system.

Design of the valley gutter drainage of a building in Portsmouth
The layout of the building and the valley gutter cross-section are shown in Figures C1 and C2. As can be seen from Figure C1, the location of the outlets has been previously defined, and these are positioned at the two ends of the gutter. The design life of the building is 20 years and, for rainwater purposes, the building is defined as Category 3 according to BS EN 12056-3. The design is for a two-minute storm duration.

The design follows guidance given in BS EN 12056-3:2000 *Gravity drainage systems inside buildings. Roof drainage, layout and calculation.*

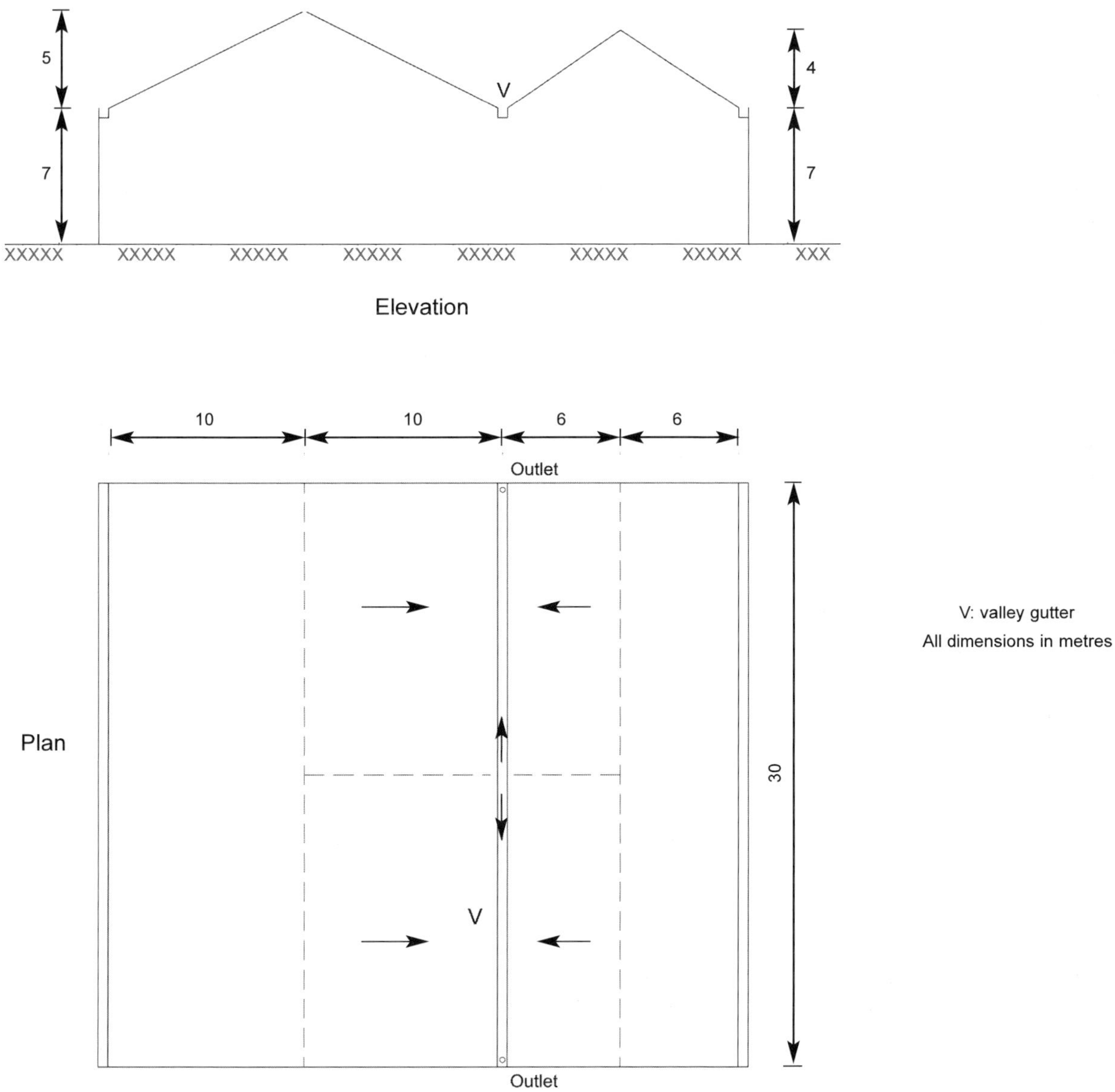

Figure C1 *Roof drainage, cross-section*

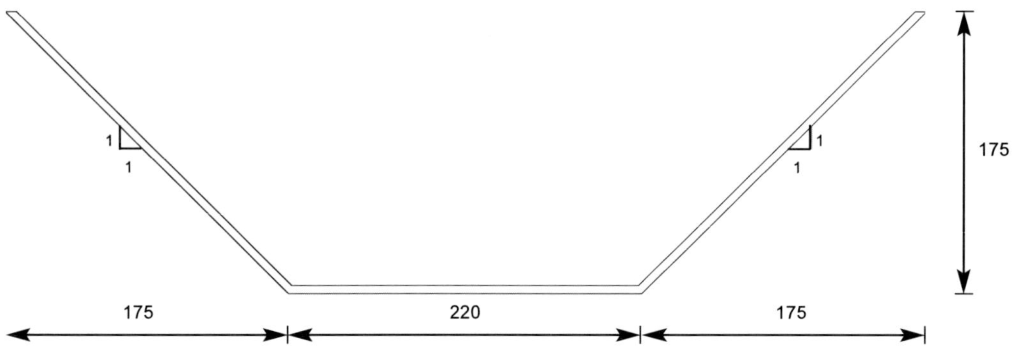

Figure C2 *Roof drainage, valley section*

C1.1 Determination of rainfall, catchment area and flow loads

Statistical rainfall data is available for the UK and is presented in National Annex NB of BS EN 12056-3. This information should be used for design.

The design storm return period factor, T, is given by:

$T = 4.5$

Therefore the design storm return period is:

$L_y = 4.5 \times 20$

$= 90$ years

Find the design rainfall intensity, r, for a two-minute duration storm with a return period of 90 years.

From Annex NB and Figure NB.2, the 2min M5 rainfall depth is 3.5 mm.

From Figure NB.7, for a return period of 90 years:

Rainfall ration = (2min M90)/(2min M5)

= 1.7

Therefore:

2min M90 $= 1.7 \times 3.5$

= 5.95 mm

The design rainfall intensity, r, is given by:

Rainfall "r" $= (2\text{min M90}) / (2 \times 60)$

$= 5.95 / (2 \times 60)$

$= 0.0495$ l/s/m^2

Therefore use r = 0.050 l/s/m^2

The catchment area can be calculated using Annex NC.

See Figure C1 for location of valley gutter.

Total plan area drained by gutter	$A_h = 30 \times 16$	$= 480\ m^2$
Total elevation area	$A_v = 30 \times (5 - 4)$	$= 30\ m^2$
Effective catchment area	$A = A_h + 0.5A_v$	$= 495\ m^2$

The rainwater runoff is given by:

$$Q = r \,.\, A \,.\, C$$
$$= 0.050 \times 495$$
$$= 24.75\ l/s$$

C1.2 Gutter design (National Annex ND)

Refer to Figure C1 for the gutter geometry (cross-section) and also to Figure C2.

Total flow in gutter = 24.75 l/s.

The outlets are at the ends of gutter, therefore gutter length:

$$GL = 30/2$$
$$= 15\ m$$

Flow load in each gutter length

$$FL = 24.75 / 2$$
$$= 12.4\ l/s$$

Flow capacity of gutter

From Table 5 determine minimum freeboard as $0.3 \times Z$ (in mm), with Z equal to depth of gutter including freeboard. Take freeboard as = 53 mm.

Maximum depth in the gutter, W is given by:

$$W = 175 - 53$$
$$= 122\ mm$$

The flow capacity of the gutter can be determined from Section 5.2.3.

$$QL = 0.9\ Q_N$$

$$Q_{SV} = 3.89 \times 10^{-5} \times [(220 + 464) \times (122/2)]^{1.25}$$
$$= 23.20\ l/s$$

$$W/T = 122/464$$
$$= 0.26$$

Therefore:

$$F_d = (W/T)^{0.25}$$
$$= 0.71$$

$$S/T = 220 / [220 + 2 \times (122)]$$
$$= 0.47.$$

Therefore:

F_s $= 0.97$

Q_N $= 23.20 \times 0.71 \times 0.97$

$= 15.98$ l/s

Q_L $= 0.9 \times 15.98$

$= 14.38$ l/s.

Check whether gutter is hydraulically "short", ie $GL \leq 50$ W:

$15\ 000 < 50 \times 122 = 6100$? False

$\therefore$ gutter is hydraulically long.

From Table 6, calculate capacity factor F_L:

L/W $= 15\ 000 / 122$

$= 123$.

For nominally level gutter, $F_L = 0.9$.

Apply F_L to the capacity of the gutter:

Q $= 0.9 \times 14.38$

$= 12.94$ l/s.

Determine whether the gutter discharges freely.

From Annex ND3.2 and Figure 10:

S/T $= 0.47$

F_h $= 0.52$.

The maximum head for free discharge, h_1, is given by:

h_1 $= F_h W$

$= 0.52 \times 122$

$= 63$ mm.

From Table 7 (assuming circular outlet, with effective diameter D = 150 mm).

Is $h_1 \leq D/2$? ie is 63 mm $\leq$ 75 mm ?

Yes $\Rightarrow$ use the Weir flow equation:

Q_o $= k_o D h^{1.5} / 7500$.

Take $k_o = 1.0$ for outlets without strainers or gratings:

Q_o $= 1.0 \times 150 \times 63^{1.5} / 7500$

$= 10.0$ l/s

$\therefore$ since Q_o (= 10 l/s) < Q (= 12.94 l/s), the outlet will not allow free discharge.

Determination of head h_R and gutter capacity for restricted discharge (Annex ND 3.3).

Take h_R - 80 mm (a value between 63 mm and 122 mm).

Calculate flow restriction factor F_R

$$F_R = (h_1 / h_R) [(W - h_R) / (W - h_1)]$$
$$= (63 / 80) [(122 - 80) / (122 - 63)]$$
$$= 0.56.$$

From Table 7:

h_R (= 80) > (150/2 = 75) → use Orifice flow equation

$$Q_{oR} = k_o D^2 h^{0.5} / 15\ 000$$
$$= 1.0 \times 150^2 \times 80^{0.5} / 15\ 000$$
$$= 13.4 \text{ l/s}.$$

From Figure ND.2, with $F_R = 0.56$:

$Q_{1R} / Q_1 = 0.86$

Since $Q_1 = 12.94$,

$$Q_{1R} = 12.94 \times 0.86$$
$$= 11.13 \text{ l/s}$$

$$Q_{1R} \neq Q_{oR}$$

Therefore, find new value for h_R and repeat the calculations until agreement is found between Q_{1R} and Q_{oR}. This was found for h_R = 72 mm as shown below:

$$F_R = (63/72) [(122 - 72) / (122 - 63)]$$
$$= 0.74$$

h_R (= 72) < (150/2 = 75), therefore use Weir flow equation:

$$Q_{oR} = 1.0 \times 150 \times 72^{1.5} / 7500$$
$$= 12.2 \text{ l/s}.$$

From Figure ND.2, with $F_R = 0.74$:

$Q_{1R} / Q_1 = 0.935$

Since $Q_1 = 12.94$

$$Q_{1R} = 12.94 \times 0.935$$
$$= 12.1 \text{ l/s}$$

Therefore, as Q_{1R} is ≈ Q_{oR} = 12.2 l/s, the design water depth in the gutter is 72 mm and the flow rate is 12.1 l/s.

C1.3 Rainwater pipe design

The rainwater pipe is assumed to be vertical, with a diameter d_i determined according to Figure 9.

If a tapered outlet is considered, the diameter of the pipe d_i is given by $D = D_o = 150$ mm $\geq 1.5d_i$. Therefore d_i needs to be ≤ 100 mm.

From Table 8 the capacity of the 100 mm-diameter pipe (assuming a filling degree of 0.33) is 10.7 l/s, which is insufficient to discharge the flow from the gutter.

Options include:

- a round-edged outlet with a top diameter of 167 mm (and $D = 150$ mm) and radius of curvature of the lip $R > 28$ mm, associated with a rainwater pipe of 110 mm internal diameter, or
- a sharp-edged outlet and a downpipe of diameter 150 mm.

D STORMWATER STORAGE ANALYSIS EXAMPLES

The following two examples, Oundle and Hinckley, illustrate the approach for designing stormwater storage where it is needed to satisfy a limiting discharge consent.

The examples use standard computer models using fixed discharge values for the limiting discharge requirements. It should be noted that the approximate methods normally employed for a first pass, using models, are often not conservative (in terms of storage volume), as the maximum flow rate for a limiting discharge is usually assumed to be effective at all water levels and does not take into account the head-discharge relationship of outflow structures.

It is stressed that, for the purpose of illustration, the example assumes the use of tanks, throttles and overflow controls. In practice, the various SUDS units would be chosen for their characteristics and their appropriateness for the soil type in the catchment so as to achieve a similar effect and to meet the flow control criteria. Figures in the examples illustrate how the model has been built and used.

D1 Structure of Appendix D

This section is in two parts. The same two examples are used to illustrate both the current methodology that is generally used and the proposed methodology produced by recent research carried out at HR Wallingford. The first example catchment, Oundle, has SOIL type 1 and the second, Hinckley, has SOIL type 4.

On the basis that the one-year greenfield runoff is used as determining the throttle limit (for the current method of analysis), this results in 0.3 l/s/ha and 3 l/s/ha respectively for the two catchments. In the case of Oundle, the throttle rate is increased to 1 l/s/ha as the site is assumed to be 10 ha in size, applying a minimum outflow rate of 10 l/s. This is needed for practical drainage reasons, as throttles below 10 l/s are generally considered to be at risk of blockage.

Many regions would regard these rates as draconian, however, so two additional typical throttle rates have also been run to illustrate storage requirements for limiting the 1 in 100-year event.

D2 Comparison of results from existing and proposed procedures

Table D1 compares the storage volumes determined by the current and proposed methods using the one-year greenfield runoff rate. Table D2 shows the other storage volumes for alternative throttle limits for the current procedure.

Table D1 shows a comparison of the existing and proposed methods. The proposed method effectively splits the storage into several parts. It is not expected that all of the storage be in the form of ponds or tanks; some could be temporary storage utilising public open space.

In both cases, the river flood protection storage is based on the Wallingford Procedure runoff model and the FSSR 16 (IH, 1985) runoff equation. The value of T_p has been calculated for the catchments (20 hours and 16 hours, respectively). These durations are longer than the default duration for "long-term" storage (six hours) and therefore the calculated storage volumes are comparatively larger as a result.

In the case of Oundle, the traditional method results in a storage volume of 2396 m^3. The proposed methodology suggests that 2420 m^3 is needed, but 1540 m^3 of this would be "permanent" or "long-term" storage and is unlikely to be provided in the form of a tank or storage pond. The volume required to protect the site due to limiting discharge constraints is only 980 m^3. This is likely to produce cost savings, depending on the mechanism used for storing the river flood protection storage.

There is no need for storage for the "level of service" criterion because of the need to mobilise as much "river flood protection" storage as possible. This flood protection is shown to be more than sufficient to cater for both the "level of service" storage and "site flood protection" storage.

In the case of Hinckley, the traditional approach results in a slightly smaller volume (than Oundle) needing 1873 m^3 and the proposed method requiring even less at 1310 m^3. Of this 1310 m^3, "site flood protection" requires 470 m^3 and this is unlikely to be provided in the form of a pond or tank. Thus only 840 m^3 would be catered for specifically to meet limiting discharge requirements, resulting in a saving of over two times that of the traditional approach. It should be noted that all these numbers are dependent on parameters and equations being selected, many of which have proposed alternatives (for example, 12 hours might be used instead of T_p), and these would affect the results. However, the principles are clearly demonstrated by these two examples of the flexibility provided by the proposed method.

Table D1 *Comparison of current and proposed methods of analysis*

Catchment	Current method		Proposed method					
	Greenfield runoff throttle rate (l/s/ha) 1	3	Water quality	River regime	Level of service	River flood protection	Site flood protection	Total
Oundle	2396	–	120	760	–	1540	–	2420
Hinckley	–	1873	120	420	300	–	470	1310

Table D2 clearly illustrates the difference in storage volumes needed for a range of discharge limits. However, although these throttles all protect the receiving water from flushing effects of "instantaneous" runoff, they generally provide little benefit in terms of river flood protection or water quality.

Table D2 *Current method of analysis – various throttle rates for 100-year critical duration events*

Catchment	Throttle rates (l/s/ha)			
	1	3	5	7
Oundle	2396	1760	1476	1300
Hinckley	2929	1873	1536	1343

D3 General comments on the use of models

The New variable PR Wallingford Procedure equation (HR Wallingford and IH, 1981) has been used. It is:

$$PR = IF \times PIMP + (100 - IF \times PIMP) \times \frac{NAPI}{PF}$$

NAPI increases with rainfall depth during the event, and therefore PR also increases. A design value for NAPI has been taken as zero. PF is the default value of 200 mm.

More information on the Old fixed PR Wallingford Procedure equation and New variable PR Wallingford Procedure equation can be found in Chapter 9.

1. **Use of hydrodynamic models**

   When modelling to determine the approximate storage required, the pipe system is often modelled with a limit of discharge throttle and an overflow. The volume passing over the overflow is the storage needed. A range of storm durations is used to determine the maximum volume.

   When detailed design (final design) of storage is carried out, models should be built that explicitly represent all storage volumes and also the head-discharge relationship of all throttle units.

D4 Comments relating to the examples used for storage assessment

2. Storage requirements

When analysing for the storage needed for a certain criterion (say "level of service" for the site), the previously defined storage (say "river regime protection") is modelled explicitly as a storage structure and an overflow weir is used to determine the additional storage needed.

3. Mobilising "river flood protection"

The volumetric analysis for "river flood protection" is not affected by design event profile and is purely a comparison of pre- and post-development runoff volumes. This means that the PR equation is used to obtain the post-development runoff volume. Although NAPI increases with rainfall during the event, this is not a linear process, as there is a decay function in the formula. If an approximation for this mean value of NAPI is needed for a manual assessment of runoff volume, a conservative assumption would be to assume that the NAPI value at the end of the event was equal to the rainfall depth added to the value of NAPI at the start of the event and therefore the mean value of NAPI could be used in a calculation of PR.

The long-term storage for "river flood protection" will not necessarily be adequately mobilised after storage has been utilised for the 20-year or 30-year level of service for the site due to the lower intensities of longer-duration rainfall. Therefore, long-term storage may need to be designed to start coming into effect at a lower return period if a system of overflow structures are being used. In practice, the use of various SUDS units would be used to more easily achieve the retention of "long-term" storage.

It is advised that time series rainfall, if it exists, is used to check that "long-term" storage is mobilised effectively. This is because volumetric aspects of storage analysis are not perfectly represented by standard design event profiles that are primarily aimed at flow rates for pipe design. However, design events do provide a good approximation for initial design.

4. Definition of the "time to peak" (T_p)

The estimation of the time to peak (T_p) has been carried out following the procedure described in the *Flood estimation handbook* (CEH, 1999). The procedure uses the unit hydrograph theory. The time to peak of the instantaneous unit hydrograph is $T_p(0)$ and can be evaluated from catchment descriptors.

$$T_p(0)=4.270\ DPSBAR^{-0.35}\ PROPWET^{-0.80}\ DPLBAR^{0.54}\ (1+URBEXT)^{-0.77}$$

T_p has been selected as an approximation for the critical time to peak of the catchment.

Examples part 1 – current procedure

The following two examples are illustrative of the current requirements for providing storage using a 100-year event. A range of results is provided to illustrate the effect of using different throttle rates on the predicted storage volumes.

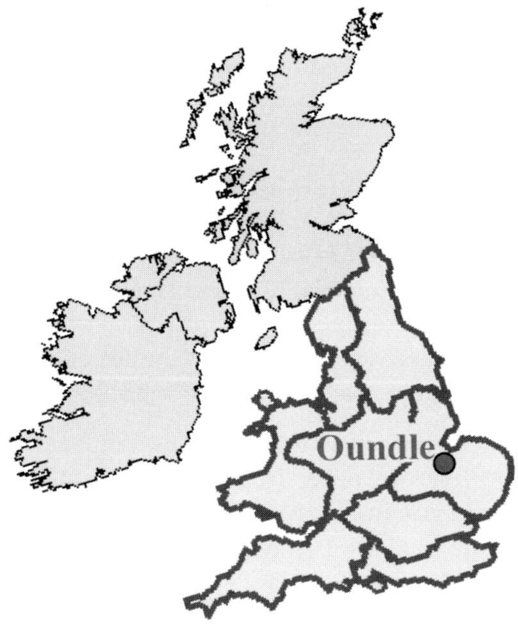

The following example illustrates the method of storage design currently practised. The theoretical catchment chosen is assumed to be at Oundle. Its characteristics are reported below together with a range of possible discharge limits to demonstrate the effect on storage volume.

CATCHMENT CHARACTERISTICS (Oundle – Anglia Region)

Site area	= 10 ha
SAAR	= 616 mm
SOIL	= 1
M5-60	= 19 mm
r	= 0.42
PIMP	= 50 per cent
IF	= 0.7
PF	= 200 mm
Throttle limits	= 1, 3, 5, 7 l/s/ha

CURRENT PROCEDURE (OUNDLE)

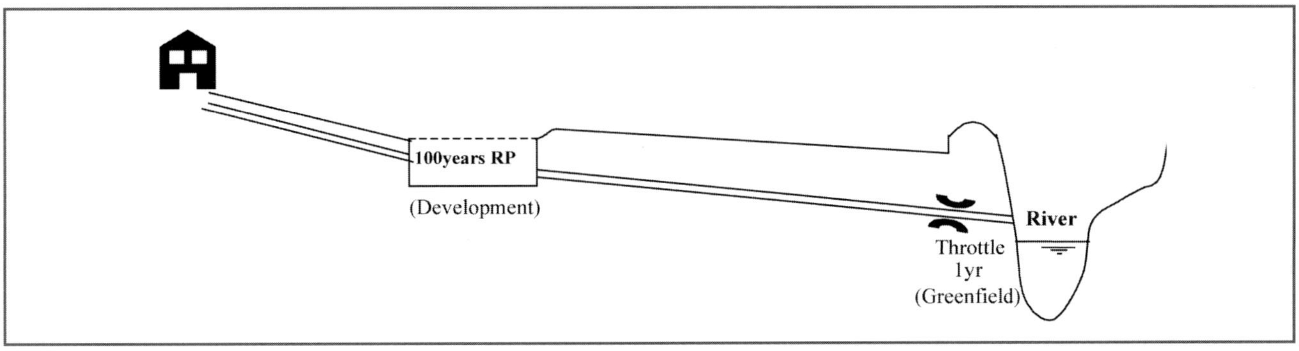

DRAINAGE MODEL

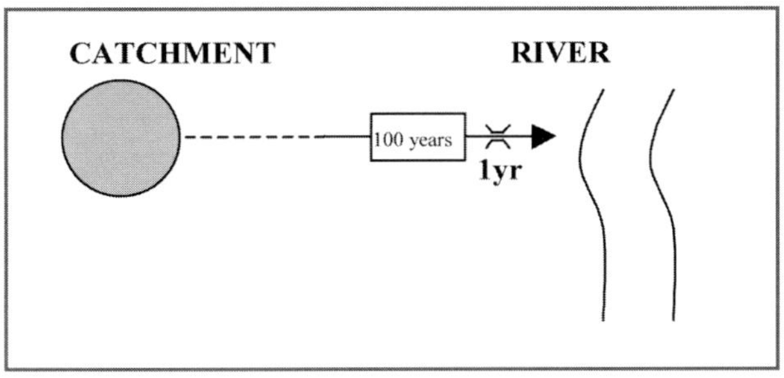

Catchment protection requires:

Throttle limit discharge = **1** yr greenfield runoff (minimum **10** l/s)

Using IH Report 124:

$Q_{rural} = 0.00108\ AREA^{0.89}\ SAAR^{1.17}\ SOIL^{2.17}$

$\mathbf{Q} = Q_{rural} \times F$

Regional growth factor (F) = 0.70

Oundle limit discharge

$\mathbf{Q} = 0.003\ m^3/s < 0.\ 010\ m^3/s$

therefore:

$\mathbf{Q} = 0.010\ m^3/s$ (1 l/s/ha)

Model run with 1 in 100-year rainfall events:

$V_{1\ l/s/ha}$ = 2396 m^3
Critical duration = 20 h

$V_{3\ l/s/ha}$ = 1760 m^3
Critical duration = 5 h

$V_{5\ l/s/ha}$ = 1476 m^3
Critical duration = 3 h

$V_{7\ l/s/ha}$ = 1300 m^3
Critical duration = 3 h

Comments

Theoretically the one-year greenfield runoff rate is only 3 l/s for this 10 ha site. 10 l/s is assumed based on practical consideration of throttle sizes. This represents 1 l/s/ha, which is generally considered as being onerous. Higher throttle rates have been provided for illustration as typical values are often in the range of 5–7 l/s/ha at present.

HINCKLEY (Midlands Region)

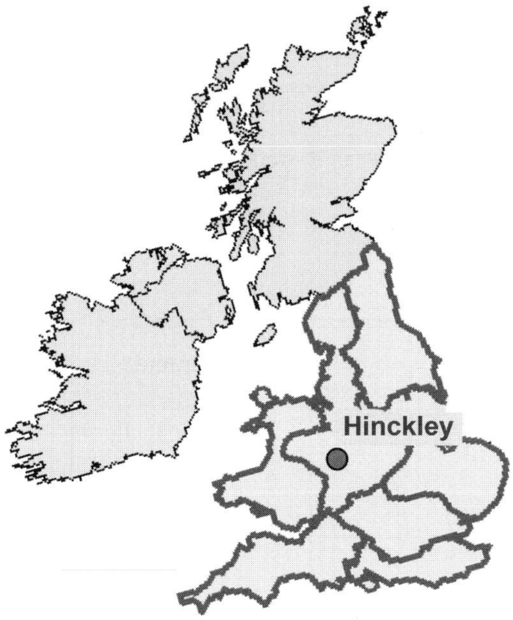

The following example illustrates the method of the storage design currently practised. The theoretical catchment chosen is assumed to be at Hinckley. Its characteristics are reported below together with a range of possible discharge limits to demonstrate the effect on storage volume.

CATCHMENT CHARACTERISTICS (Hinckley – Midland Region)

Site area	= 10 ha
SAAR	= 682 mm
SOIL	= 4
M5-60	= 19 mm
R	= 0.40
PIMP	= 50 per cent
IF	= 0.7
PF	= 200 mm
Throttle limits	= 1, 3, 5, 7 l/s/ha

CURRENT PROCEDURE (HINCKLEY)

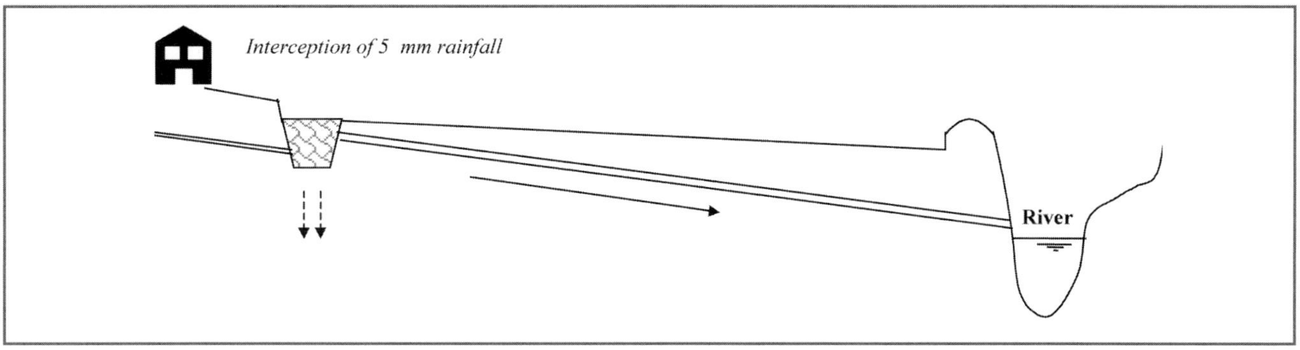

DRAINAGE MODEL

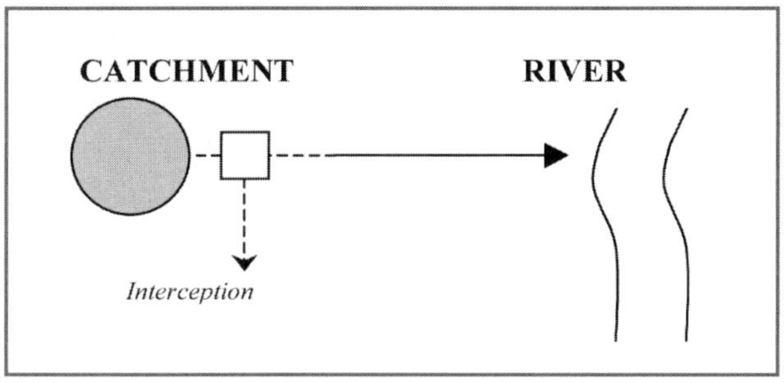

Catchment protection requires:

Throttle limit discharge = **1** yr greenfield runoff (minimum **10** l/s)

Using IH Report 124:

$Q_{rural} = 0.00108\ AREA^{0.89}\ SAAR^{1.17}\ SOIL^{2.17}$
$= 0.048\ m^3/s$

$\mathbf{Q} = Q_{rural} \times F$

Regional growth factor (F) = 0.65

Hinckley limit discharge = Q

$\mathbf{Q} = 0.030\ m^3/s > 0.010\ m^3/s$

Model run with 1 in 100-year rainfall event:

$V_{1\ l/s/ha}$ = 2829 m^3
Critical duration = 33 h

$V_{3\ l/s/ha}$ = 1873 m^3
Critical duration = 7 h

$V_{5\ l/s/ha}$ = 1536 m^3
Critical duration = 5 h

$V_{7\ l/s/ha}$ = 1343 m^3
Critical duration = 3 h

Comments

Examples part 2 – proposed procedure

The following examples are illustrations of the proposed storage methodology. This is more complex than the simple throttle and storage approach currently used and illustrated in Part 1, but in principle the modelling processes used are the same.

Note

For clarification, figures carried forward from calculations made earlier on in the example appear in grey type (as below).

V interception = 120 m^3

OUNDLE (Anglia Region)

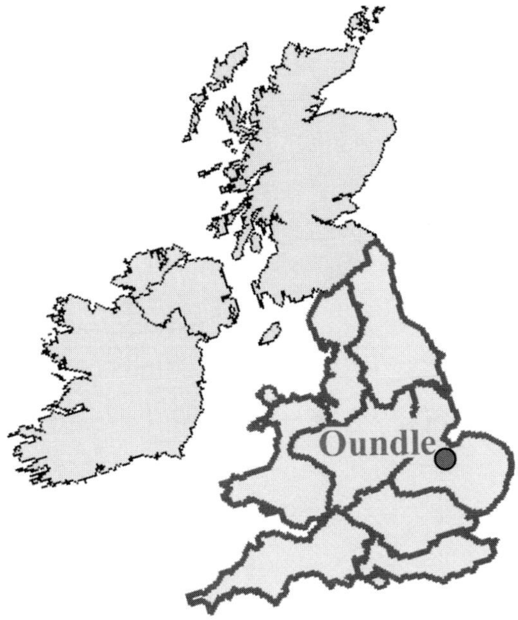

The following example illustrates the use of the proposed storage design procedure. The theoretical catchment chosen is assumed to be at Oundle. Its characteristics are reported below together with requirements for each of the criteria considered.

CATCHMENT CHARACTERISTICS (Oundle – Anglia Region)

Site area	= 10 ha
SAAR	= 616 mm
SOIL	= 1
M5-60	= 19 mm
r	= 0.42
PIMP	= 50 per cent
IF	= 0.7
PF	= 200 mm
Catchment area	= 90 km^2
Receiving river	= River Willow Brook
Catchment T_p	= 20.1 h

Criterion	Rainfall interception	Runoff limit discharge	"Attenuation" storage event	"Long-term" storage event	"Flood" storage event
1 River water quality protection	5 mm (min 20 m^3)	–	–	–	–
2 River regime protection	–	1-year RP site greenfield runoff (min 10 l/s)	5-year RP (site-critical duration)	–	–
3 Level of service for the site	–	20-year RP site greenfield runoff (min 10 l/s)	20-year RP (site-critical duration)	–	–
4 River flood protection	–	–	–	100-year RP (T_p catchment-critical duration)	–
5 Site flood proteection	–	100-year RP site greenfield runoff (min 10 l/s)	–	–	100-year RP (site-critical duration)
6 Site catastrophic protection	–	–	–	–	–

The figure below shows the analysis path followed by the example and the layout of the numerical model used for preliminary estimation of storage requirements.

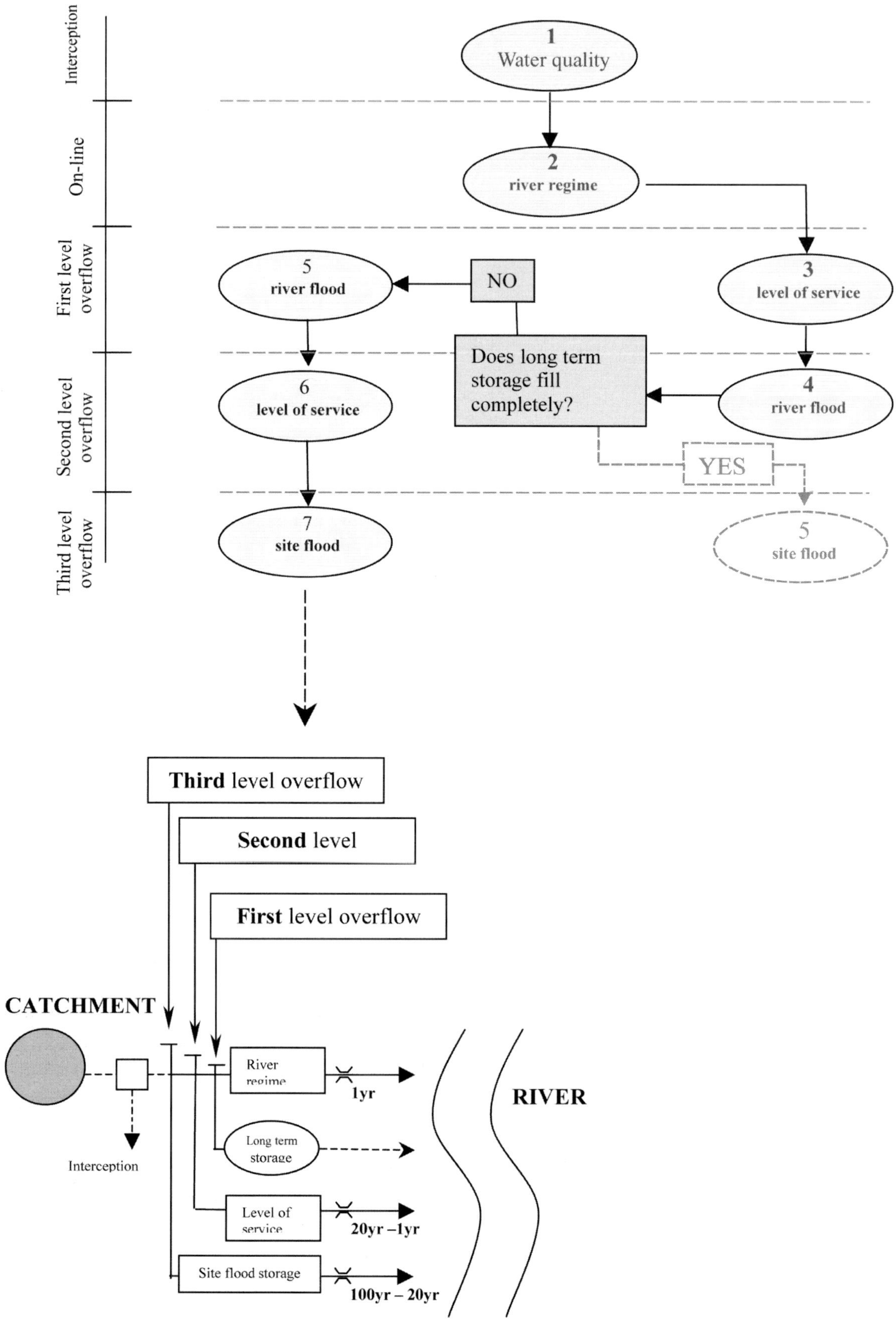

PROPOSED PROCEDURE

1. WATER QUALITY PROTECTION (OUNDLE)

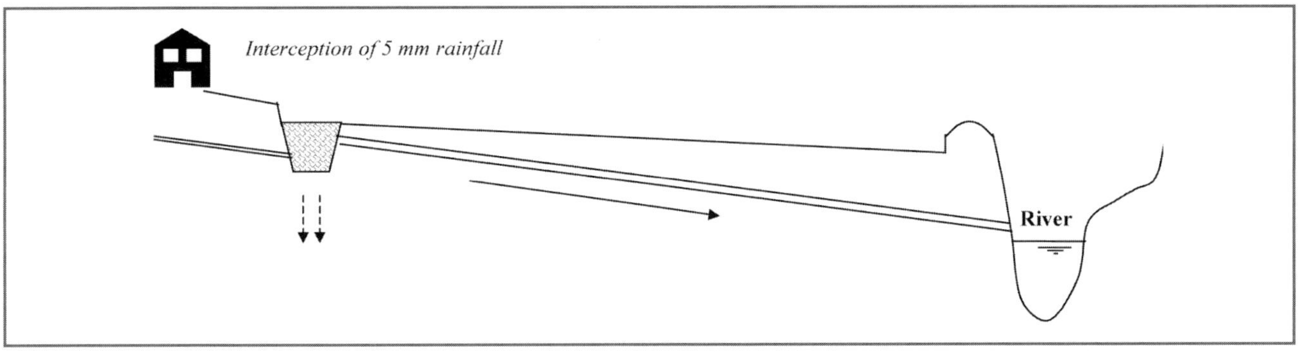

DRAINAGE MODEL

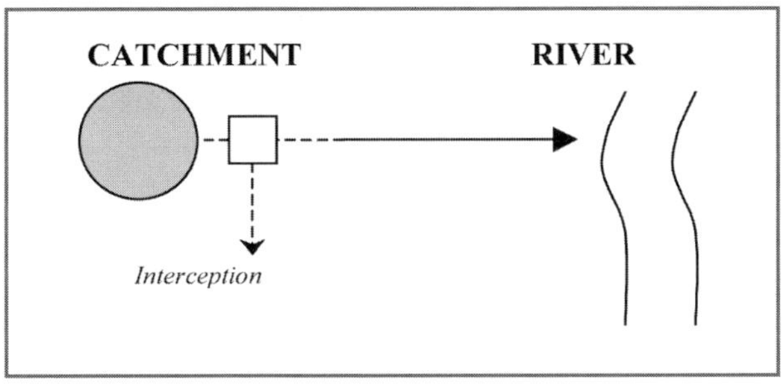

The volume needed to intercept 5 mm of rainfall is:

$V_{\textbf{interception}}$ = 120 m³ i

Model run selected year from TSR (1993) ii

TSR characteristics:
Minimum rainfall depth = 2 mm iii
Minimum antecedent dry period = 6 h

Events causing runoff: iv

Interception	No of discharges to river	
	Summer	All year
0 mm	20	56
5 mm	9	28

Comments

The approximate volume needed to intercept 5 mm of rainfall can be manually evaluated as follows:

$$V_{\textbf{interception}} = \text{Area} \times \text{PIMP} \times \text{IF} \times \text{Rainfall depth}$$
$$= 100\,000 \times 0.5 \times 0.7 \times 0.005$$
$$= 175 \text{ m}^3$$

This assumes no runoff from permeable areas during the first 5 mm of rainfall and also no depression storage on paved surfaces.

Notes

i Modelling method used:
  a) create rainfall event of 5 mm
  b) standard values used for depression storage.

ii Selection of a typical year could be improved upon by selecting 10 consecutive years. Real data (rather than statistical) should be used if it exists.

iii Time series analysis assumptions to assess number of events:
  a) any rainfall with less than 2 mm of rainfall results in minimal runoff
  b) any rainfall taking place within six hours of previous rainfall is considered to be the same event
  c) summer is the period May to September inclusive.

iv Analysis of rainfall for number of events before and after subtracting 5 mm depth of rain.

PROPOSED PROCEDURE

2. RIVER REGIME PROTECTION (OUNDLE)

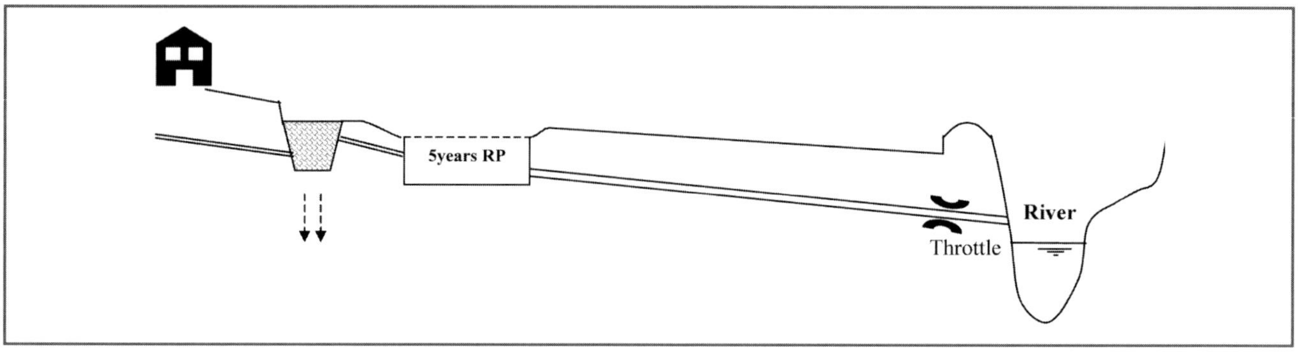

DRAINAGE MODEL

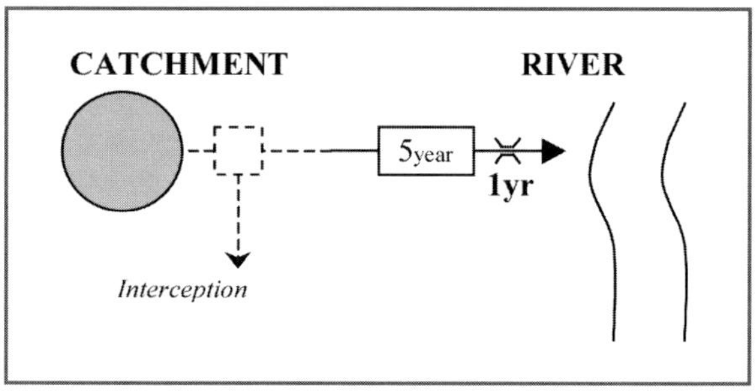

River regime protection requires:

Throttle limit discharge = **1** yr greenfield runoff (minimum **10** l/s)

Using IH Report 124:

$Q_{rural} = 0.00108\ AREA^{0.89}\ SAAR^{1.17}\ SOIL^{2.17}$

$\mathbf{Q} = Q_{rural} \times F$

Regional growth factor (F) = 0.70

Oundle limit discharge

$\mathbf{Q} = 0.003\ m^3/s < 0.010\ m^3/s$

therefore:

$\mathbf{Q} = 0.010\ m^3/s$

Model run with 1 in 5-year rainfall events. The critical duration is found to be 12 h:

$\mathbf{Vol}_{\text{interception + 5yr}}$ = 880 m^3 v

As:

$\mathbf{Vol}_{\text{interception}}$ = 120 m^3

therefore:

$\mathbf{Vol}_{\text{5yr storage}}$ = 760 m^3

Comments

The long critical duration is due to the very tight discharge limit of 1 l/s/ha. A shorter duration event would be relevant if either a smaller event was used (one year, say) or a larger throttle limit.

Notes

v The model was run without the interception storage. Therefore river regime storage is 880–120 m^3.

PROPOSED PROCEDURE

3. LEVEL OF SERVICE FOR THE SITE (OUNDLE)

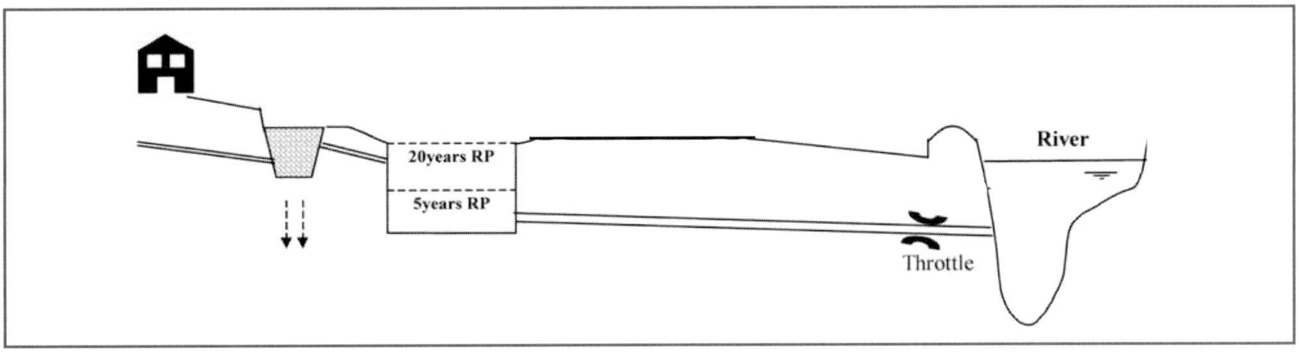

DRAINAGE MODEL

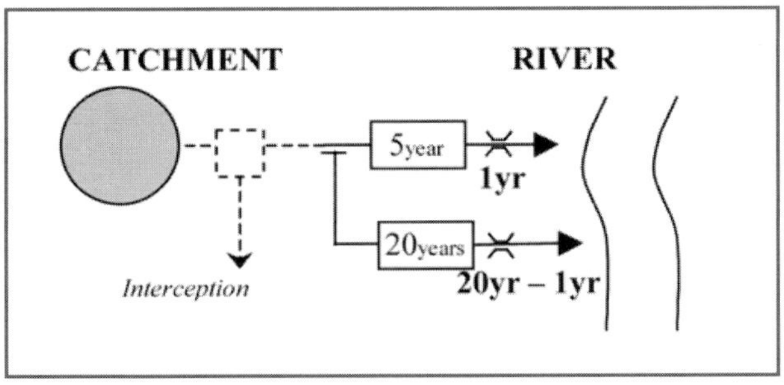

Level of service for the site requires:

Limit discharge = **20** yr greenfield runoff
(minimum **10** l/s)

Using IH Report 124:

Oundle limit discharge

Q = 0.008 m³/s < 0.010 m³/s

Therefore:

Q = 0.010 m³/s

Model run with 1 in 20-year rainfall events. The site-critical duration is found to be 15 h.

$V_{\text{interception}}$ = 120 m³
$V_{\text{5yr storage}}$ = 760 m³

As:
Vol $_{\text{interception + 5yr+20yr storage}}$ = 1480 m³

Therefore:
Vol $_{\text{20yr storage}}$ = 600 m³ vi

Comments

Notes

vi The analysis of the 5-year and 20-year volumes is carried out as a single storage structure as the throttle rate is 10 l/s for both cases. See Hinckley example, where 5-year and 20-year criteria flow rates are different.

4. RIVER FLOOD PROTECTION (OUNDLE)

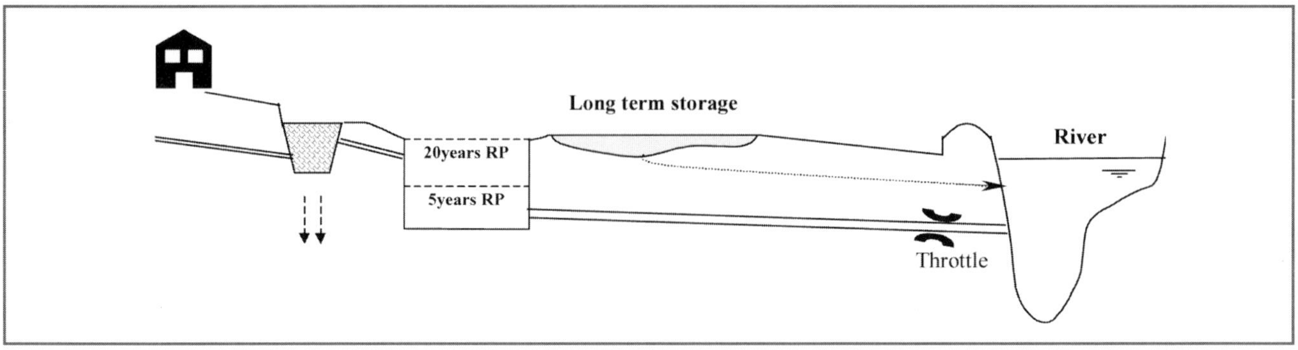

DRAINAGE MODEL

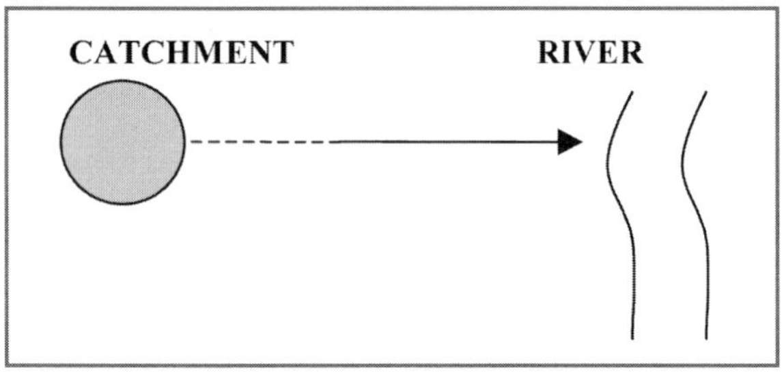

$\mathbf{V}_{\mathbf{100yr\ perm\ stor}} = \mathbf{Runoff}_{\mathbf{Develop}} - \mathbf{Runoff}_{\mathbf{Greenfield}}$

Using *Flood studies supplementary report* 16:

$PR_{Greenfield} = SPR + DPR_{CWI} + DPR_{rain}$
$= 10 + 0 + 0.45\ (76.5 - 40)^{0.7}$
$= 15.58$

$Runoff_{Greenfield} = PR_{Greenfield} \times A \times RD$ vii
$= 0.155 \times 100\ 000 \times 0.0765$
$= 1190\ m^3$

Model run with 1 in 100-year rainfall event. No structures included. Using MicroFSR the catchment duration T_p is equal to 20 h (see Appendix D4 note 5)

$Runoff_{Development} = 3176\ m^3$ (model output) viii

$\mathbf{Vol}_{\mathbf{additional\ runoff}} = 3176 - 1190 = 1986\ m^3$ ix

Model run with 1 in 100-year winter rainfall event. Catchment duration $T_p = 20$ h: x

$V_{interception} = 120\ m^3$
$V_{5yr} = 760\ m^3$
$V_{20yr} = 600\ m^3$

Spill volume from model:
$\mathbf{Vol}_{\mathbf{100yr\ long\ term\ storage}} = 980\ m^3 < 1986\ m^3$

Inadequate river flood protection: analyse "river flood protection" after "river regime protection"

Comments

Mobilising "river flood protection"

The volumetric analysis for "river flow protection" is not affected by design event profile and is purely a comparison of pre- and post-development runoff volumes. This means that the PR equation is used to determine runoff volume after development.

The "long-term" storage for "river flood protection" has not been adequately mobilised after storage has been utilised for the 20-year level of service for the site. Therefore, "long-term" storage will need to be designed to come into effect at a lower return period.

It is advised that time series rainfall be used to check that permanent storage is mobilised effectively. This is because volumetric aspects of storage analysis are not perfectly represented by standard design event profiles that are primarily aimed at pipe size analysis and flow rates.

Notes

vii A = Area (m^2) and RD = Rainfall depth (m)

viii Model run without interception storage.

ix Storage required in order to reproduce greenfield runoff.

x Model run with 5- and 20-year storage to find spill volume for storing additional greenfield runoff.

PROPOSED PROCEDURE

5. RIVER FLOOD PROTECTION (OUNDLE)

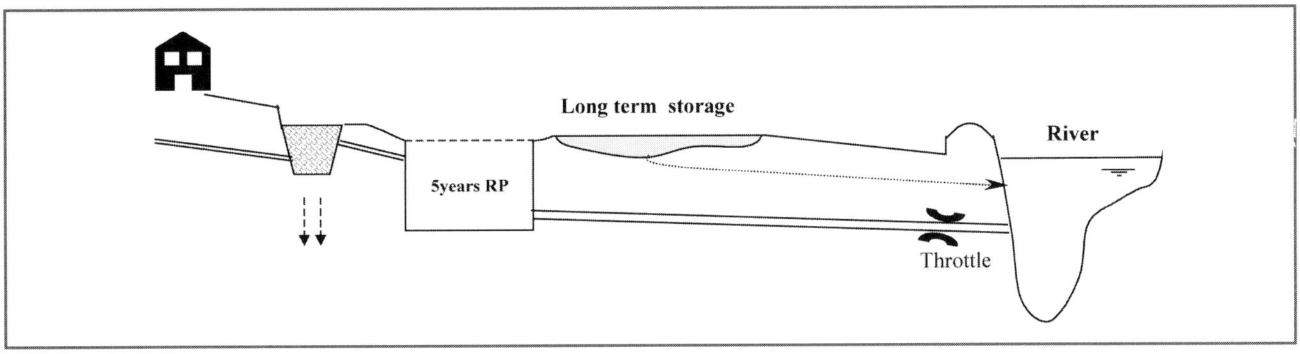

DRAINAGE MODEL

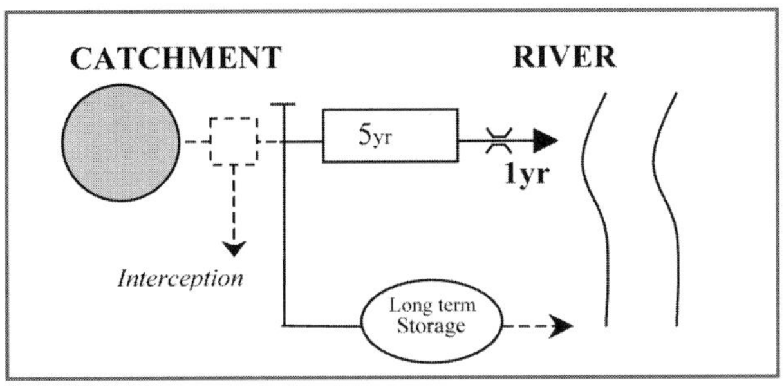

Model run with 1 in 100-year rainfall event. Catchment duration T_p = 20 h

$V_{\text{interception}}$ = 120 m^3
$V_{\text{5yr storage}}$ = 760 m^3

$V_{\text{100yr long term storage}}$ = 1540 m^3

Although 1540 m^3 < 1986 m^3 a long-term storage of 1540 m^3 is accepted. x i

Comments

1) Note that Stage 4 is now being carried out before Stage 3 as shown in the flow chart at the start of this example.
2) This result illustrates the massive increase in runoff where development sites take place on areas of SOIL type 1.

Notes

x i Long-term storage is not fully utilised due to the minimum throttle discharge rate of 10 l/s and the rainfall profile for storm duration T_p. A greater volume could only be utilised if river regime storage and interception storage were reduced. A long-term storage of 1540 m^3 is therefore accepted for river flood protection in this case. It is likely that for SOIL type 1 the criteria for long-term storage may be relaxed in some cases. The alternative is to have flooding of "long-term" storage taking place more frequently than five years

PROPOSED PROCEDURE

6. LEVEL OF SERVICE (OUNDLE)

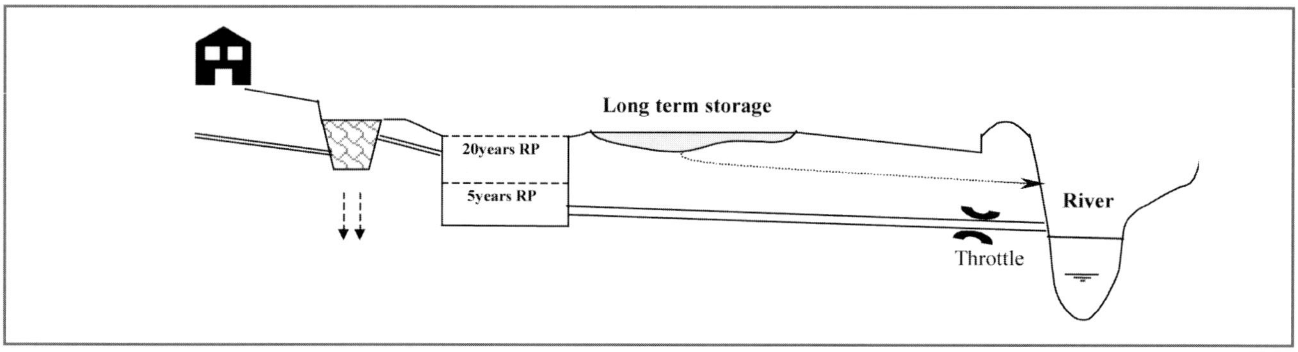

DRAINAGE MODEL

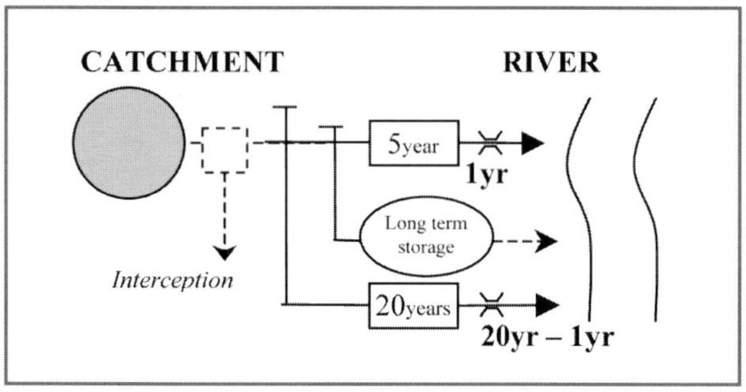

Model run with 1 in 20-year rainfall event. Site critical duration (15 h).

$V_{\text{on-line tank 5yr}}$ = 880 m^3

$\textbf{Vol}_{\text{on-line tank 20yr}}$ = 0 m^3 [xii]

$\textbf{Vol}_{\text{100yr long term storage}}$ = 550 m^3

$\textbf{Vol}_{\text{interception + 5yr+20yr+long term storage}}$ = 1430 m^3

Comments

Notes

[xii] This shows that 20-year storage is not required as the long term storage of 1540 m^3 is not filled for a return period of 20 years.

Note that throttle rate analysis is shown on level of service analysis previously

PROPOSED PROCEDURE

7. SITE FLOOD PROTECTION (OUNDLE)

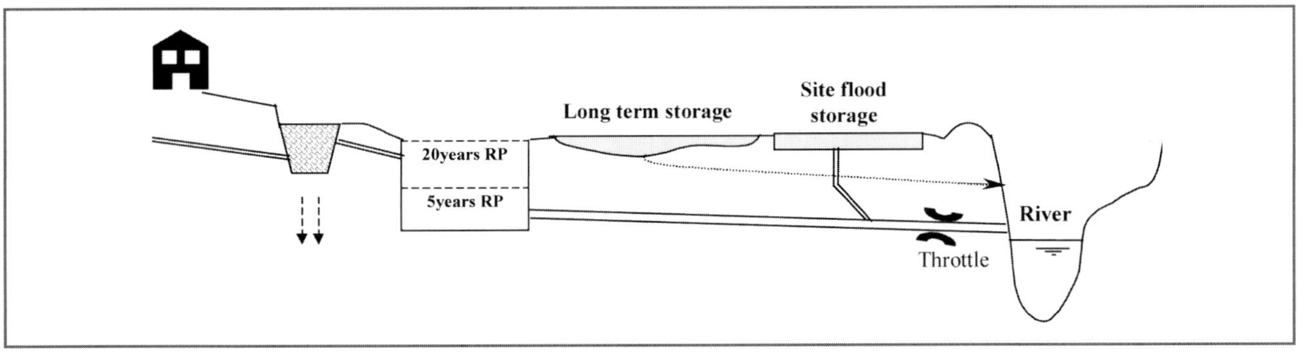

DRAINAGE MODEL

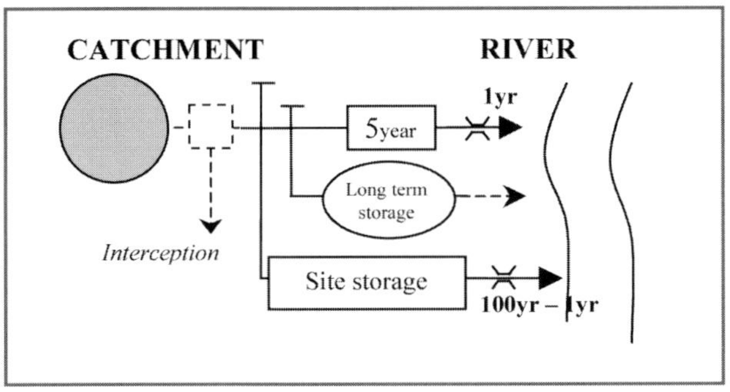

Site protection requires:
Limit discharge = **100**-year greenfield runoff
(minimum **10** l/s)

Using IH Report 124:

Oundle limit discharge **= 0.015 m³/s**

Model run with 1 in 100-year rainfall event. The site critical duration is 20 hours. [xiii]

Summarising:

Vol $_{\textbf{interception}}$ = 120 m³
Vol $_{\textbf{5yr}}$ = 760 m³
Vol $_{\textbf{20yr}}$ = 0 m³

Vol $_{\textbf{100yr long term storage}}$ = 1540 m³
Vol $_{\textbf{100yr flood storage}}$ = 0 m³

As a consequence:

Vol $_{\textbf{interception + 5yr+100yr long term storage}}$ = 2420 m³

Comments

This analysis shows that the catchment requires interception and five-year storage with all higher return period events up to 100 years all passing to "long-term" storage.

Site flood protection at the 100-year return period should cater for the effect of the 100-year limit of discharge and also short high-intensity events where local flooding around the site will take place.

Notes

xiii Site-critical duration and catchment T_p are the same.

CONCLUSION – OUNDLE

Drainage model

According to the greenfield runoff formula the 1 in 1-year discharge for Oundle is 3 l/s and the 1 in 20-year is 8 l/s. The discharge limit of 10 l/s is set as a minimum throttle discharge limit and the rainfall profile for storm duration T_p results in the long-term storage not being fully utilised.

A long-term storage of 1540 m^3 is accepted for "river flood protection" because it is not possible to mobilise more volume without mobilising "long-term" storage more frequently than a five-year return period.

The results obtained following the standard procedure can be graphically illustrated as follows. When analysing for level of service before "long-term" storage it showed that "long-term" storage is not sufficiently utilised: river flood protection is therefore not satisfied.

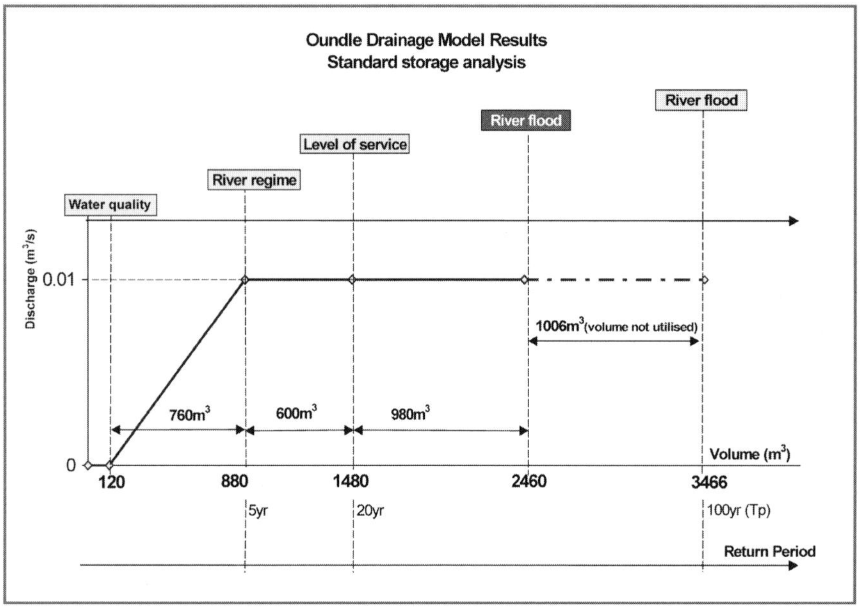

The following figure summarises the results obtained following the alternative storage analysis by running Stage 4 after Stage 2, before assessing Stages 3 and 5.

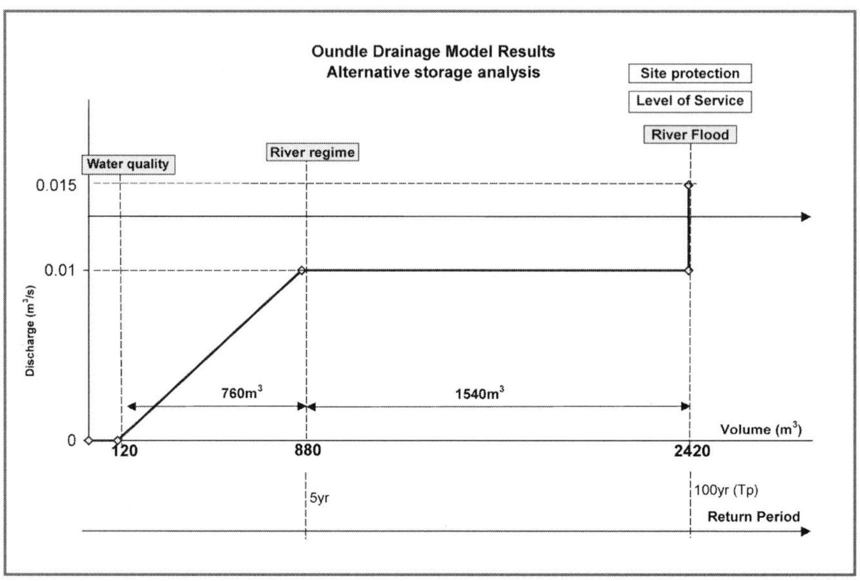

Once the long-term storage required by "river flood protection" (1540 m³) is in place, it also provides the storage required by "level of service" and "site flood protection". Although the total storage provided is much the same, a much larger proportion of storage is being retained for long-term retention. Although the simple model used produced these results, it would be expected that the total storage calculated would be greater where the "river flood protection" is utilised before the level of service criterion.

TSR analysis

The model was run with the Oundle 21 years time series rainfall. The results appear to satisfy the river flood protection requirements.

The 100-year long-term storage is mobilised 15 times. This is more frequent than expected because in the analysis the storage volume providing river flood protection is mobilised immediately after the storage providing river regime protection, which is designed for a return period of five years. The TSR analysis indicates that it occurs nearly every year.

Eleven out of the 15 events mobilising the 100-year long-term storage are "flooding events", which means they occurred when the flow in the river was above the Q_{10} flow rate. The following figure shows the 100-year long-term storage volumes mobilised during each of these events.

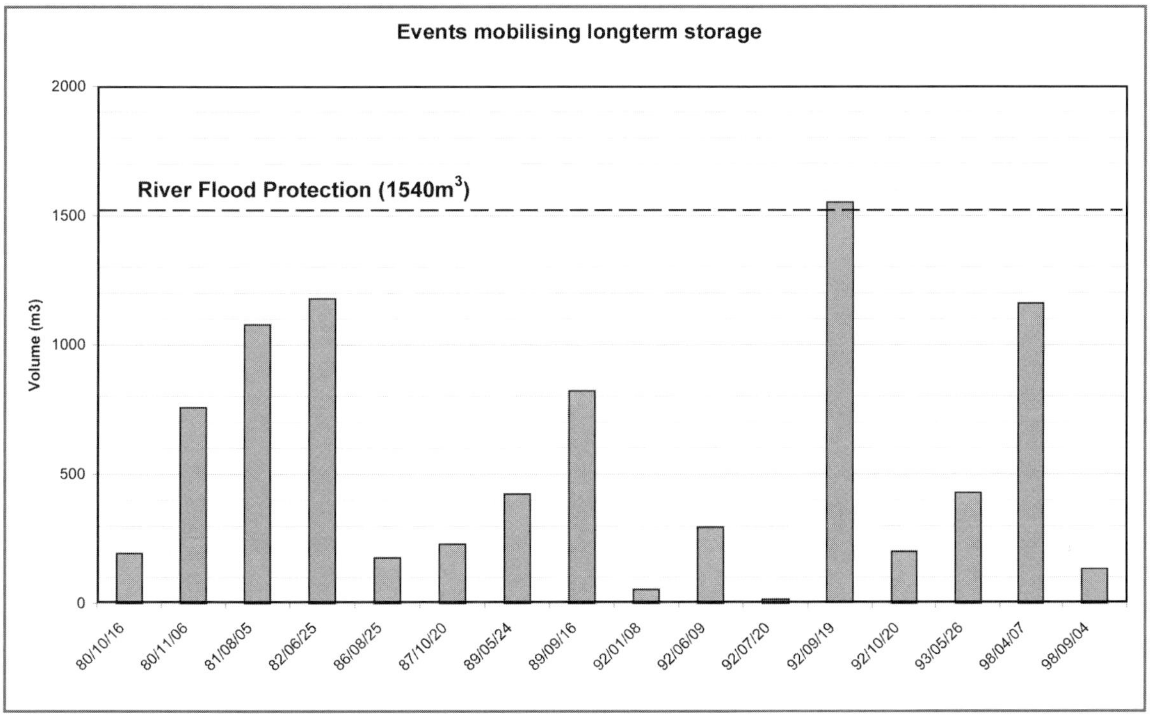

In addition to the long-term storage being mobilised, the design volume of 1540 m³ was exceeded once by the TSR events and would have mobilised the site flood protection, if any had been provided. The 100-year site flood protection storage is mobilised once with 30 m³ on 19 September 1992. The event mobilising this storage is a "river flooding event".

HINCKLEY (Midland Region)

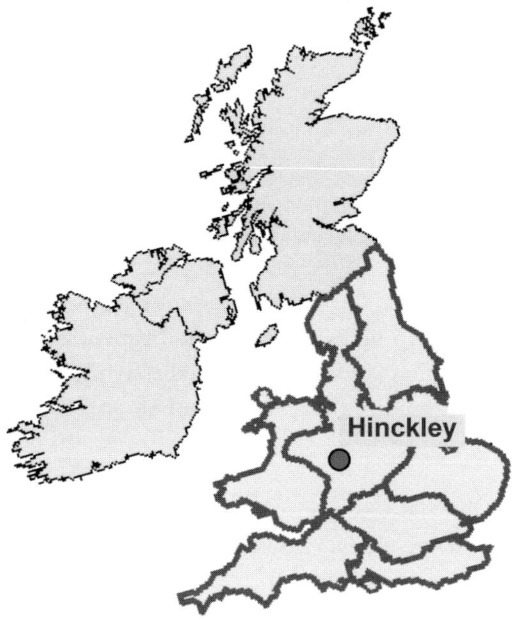

This second example illustrates the use of the proposed storage design procedure for a clay soil catchment. The theoretical catchment chosen is assumed to be at Hinckley. Its characteristics are reported below together with requirements for each of the criteria considered.

CATCHMENT CHARACTERISTICS (Hinckley – Midland Region)

Site area	= 10 ha
SAAR	= 682 mm
SOIL	= 4
M5-60	= 19 mm
R	= 0.40
PIMP	= 50 per cent
IF	= 0.7
PF	= 200 mm
Catchment area	= 262 km^2
Receiving river	= River Sowe
Catchment T_p	= 15.6 h

Criterion	Rainfall interception	Runoff limit discharge	"Attenuation" storage event	"Long-term" storage event	"Flood" storage event
1 River water quality protection	5 mm	–	–	–	–
2 River regime protection	–	1-year RP site greenfield runoff (min 10 l/s)	5-year RP (site-critical duration)	–	–
3 Level of service for the site	–	20-year RP site greenfield runoff (min 10 l/s)	20-year RP (site-critical duration)	–	–
4 River flood protection	–	–	–	100-year RP (T_p catchment-critical duration)	–
5 Site flood proteection	–	100-year RP site greenfield runoff (min 10 l/s)	–	–	100-year RP (site-critical duration)
6 Site catastrophic protection	–	–	–	–	–

The figure below shows the path followed during the examples and the layout of the numerical model use for preliminary estimation of storage requirements.

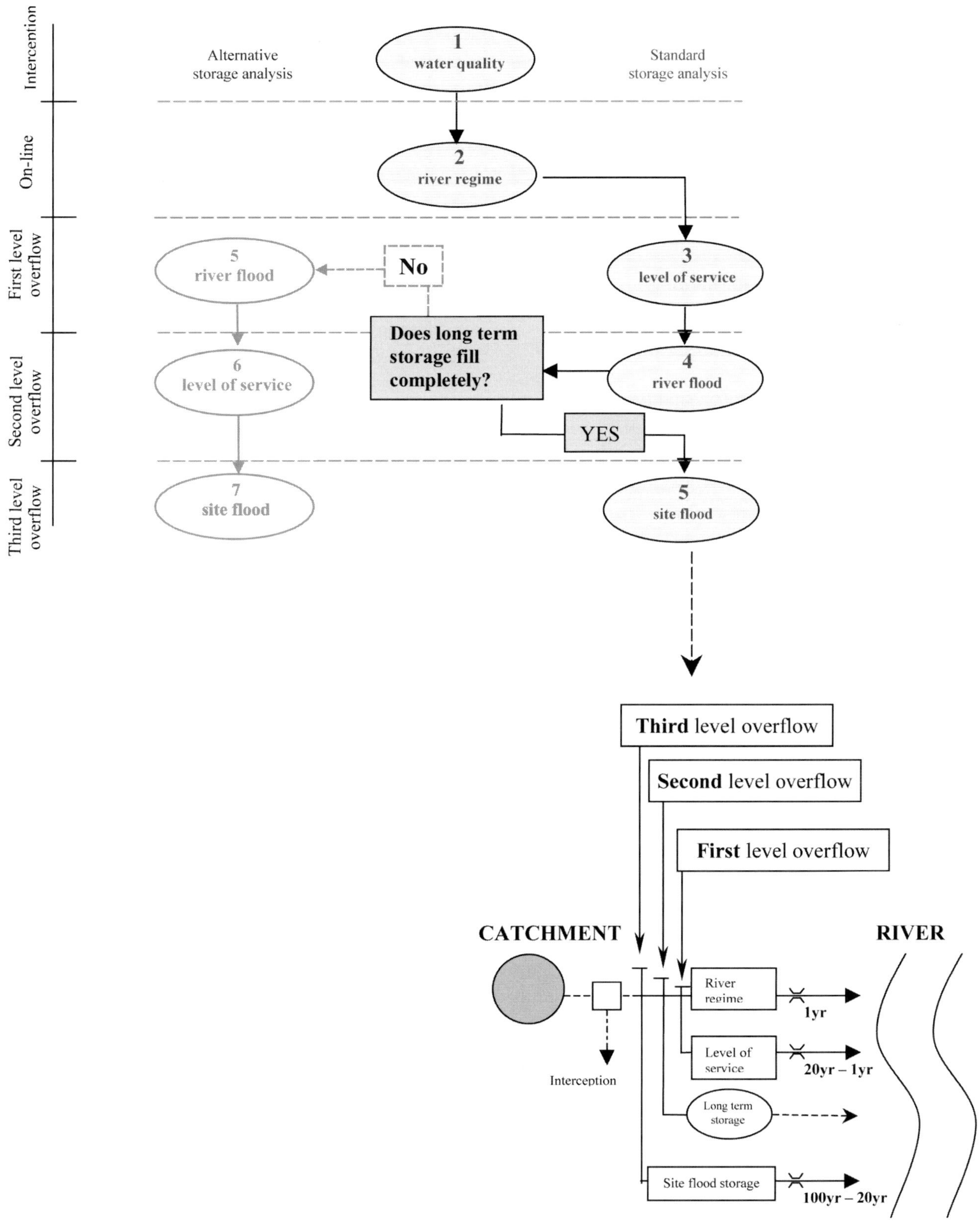

PROPOSED PROCEDURE

1. RIVER WATER QUALITY PROTECTION (HINCKLEY)

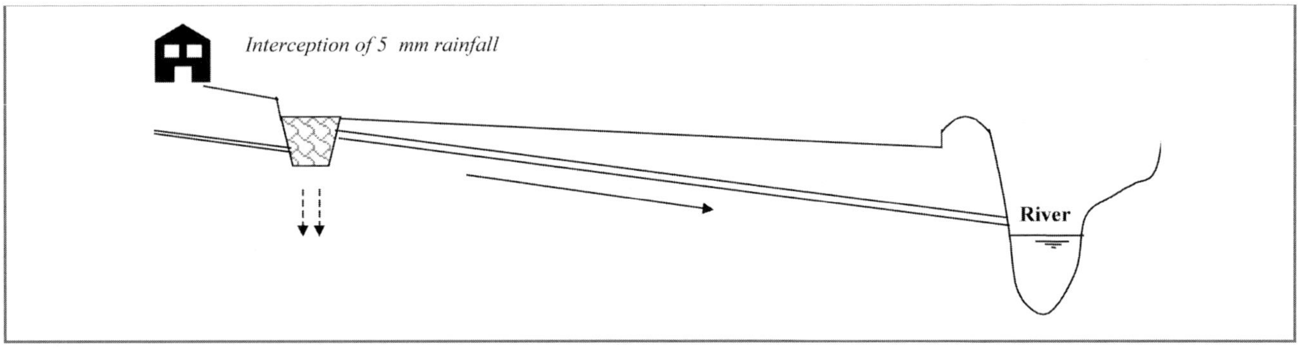

DRAINAGE MODEL

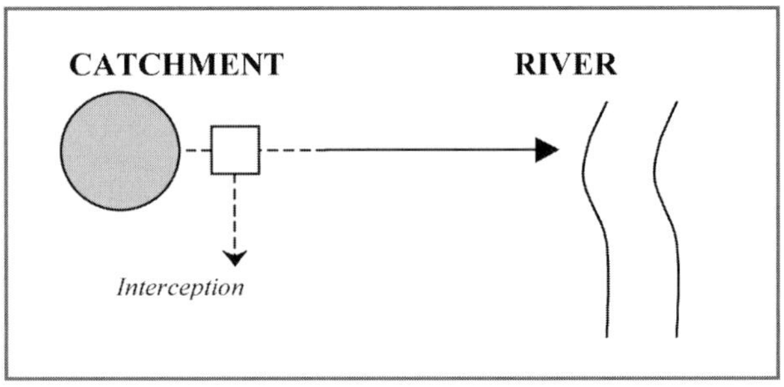

The volume needed to intercept 5 mm of rainfall is:

$V_{interception}$ = 120 m³ i

Model run selected year from TSR (1993) ii

TSR characteristics:

Minimum rainfall depth = 2 mm iii

Minimum antecedent dry period = 6 h

Events causing runoff: iv

Interception	No of discharges to river	
	Summer	All year
0 mm	25	68
5 mm	12	36

Comments

The volume needed to intercept 5 mm of rainfall can be manually evaluated as follows:

$V_{interception}$ = Area × PIMP × IF × Rainfall depth
= 100 000 × 0.5 × 0.7 × 0.005
= 175 m^3

Notes

i Modelling method used:
Standard rainfall event of 5 mm
Standard values used for depression storage

ii Selection of a typical year could be improved upon by selecting 10 consecutive years. Real data (rather than statistical) should be used.

iii Time series analysis assumptions:
Any rainfall with less than 2 mm of rainfall results in minimal runoff.
Any rainfall taking place within six hours of previous rainfall is considered to be the same event.

iv Analysis of rainfall for number of events before and after subtracting 5 mm depth of rain.
Summer is May to September inclusive.

PROPOSED PROCEDURE

2. RIVER REGIME PROTECTION (HINCKLEY)

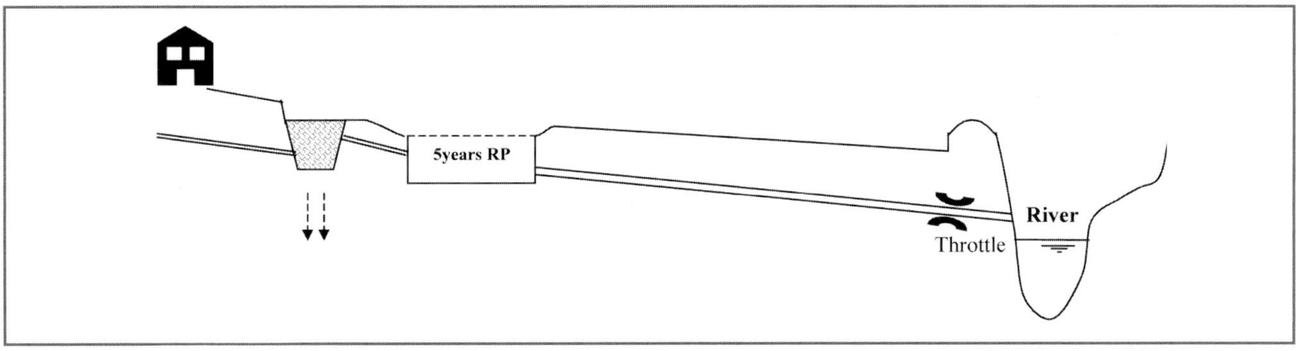

DRAINAGE MODEL

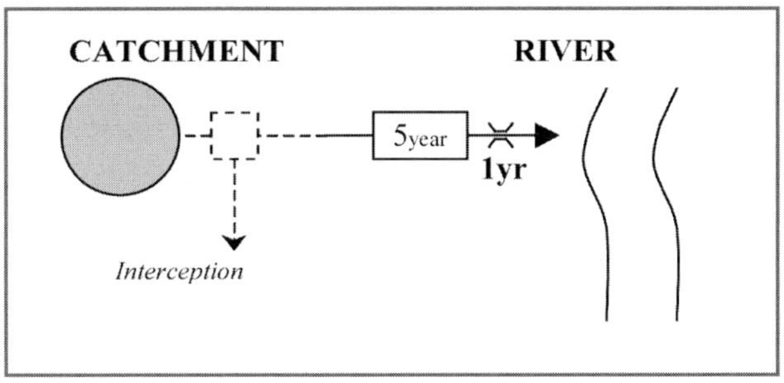

River regime protection requires:

Throttle limit discharge = **1**-year greenfield runoff (minimum **10** l/s)

Using IH Report 124:

Q_{rural} = 0.00108 $AREA^{0.89}$ $SAAR^{1.17}$ $SOIL^{2.17}$
= 0.048 m^3/s

Q = Q rural × F

Regional growth factor (F) = 0.65

Hinckley limit discharge = Q

Q = 0.030 $m^3/s > 0.010$ m^3/s

Model run with 1 in 5-year rainfall event. The site-critical duration is four hours:

$\mathbf{V}_{\textbf{interception + 5years}}$ = 540 m^3

As:

$\mathbf{V}_{\textbf{interception}}$ = 120 m^3

Therefore

$\mathbf{V}_{\textbf{5yr storage}}$ = 420 m^3

Comments

Notes

3. LEVEL OF SERVICE FOR THE SITE (HINCKLEY)

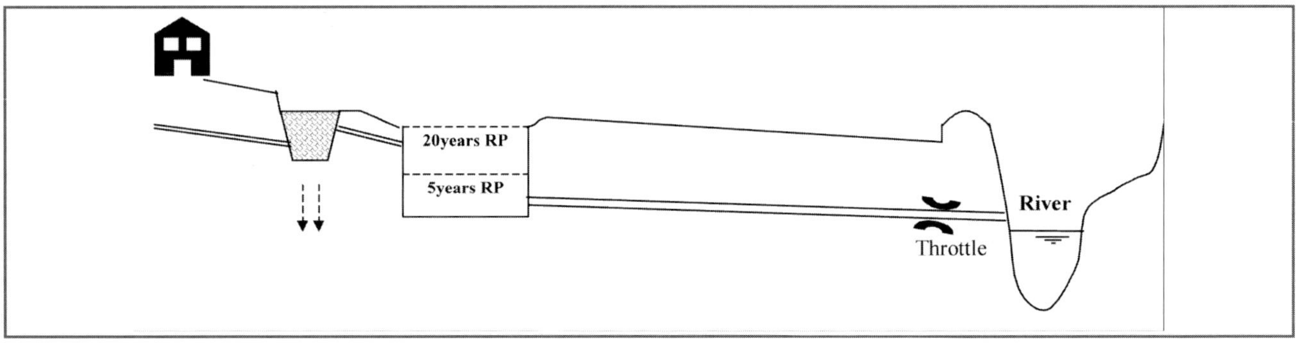

DRAINAGE MODEL

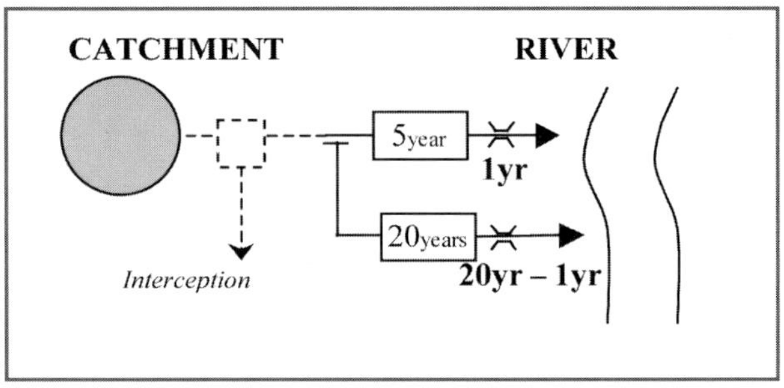

Level of service for the site requires:

Limit discharge = **20**-year greenfield runoff (minimum **10** l/s)

Using IH Report 124:

Hinckley limit discharge

Q = 0.084 $m^3/s > 0.010\ m^3/s$

therefore:

Q = 0.084 m^3/s

Model run with 1 in 20-year rainfall events. The site-critical duration is four hours.

$V_{interception}$ = 120 m^3

$V_{5yr\ storage}$ = 420 m^3

As:

$V_{interception + 5yr+20yr\ storage}$ = 840 m^3

therefore:

$V_{20yr\ storage}$ = 300 m^3 v

Comments

Notes

v The analysis of the five-year and 20-year volumes is carried out considering two storage structures connected by a weir. When the five-year storage is full (540 m^3) the water spills into the 20-year storage (300 m^3). The total discharge rate to river is 84 l/s: because 30 l/s is the discharge limit for river regime protection (see Step 2) the remaining 54 l/s represent the limit discharge from the 20-year storage structure.

PROPOSED PROCEDURE

4. RIVER FLOOD PROTECTION (HINCKLEY)

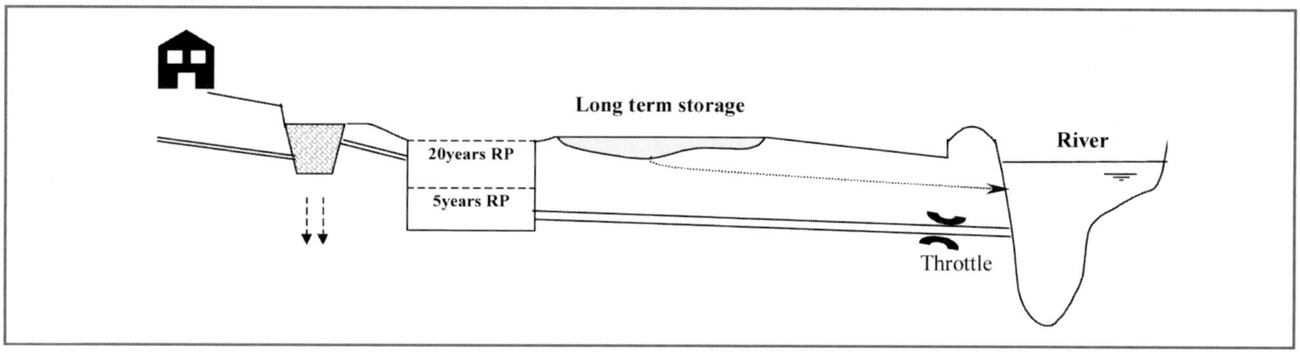

DRAINAGE MODEL

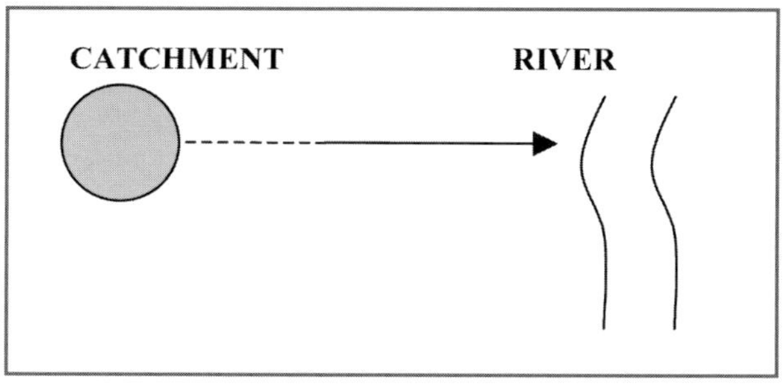

$V_{\text{100yr perm stor}}$ = **Runoff** $_{\text{Develop}}$ - **Runoff** $_{\text{Greenf}}$

Using *Flood studies supplementary report* no 16:

$PR_{\text{Greenfield}}$ = SPR + DPRCWI + PRrain

= $47 + 0 + 0.45\,(75.15 - 40)^{0.7}$

= 52.43

$\text{Runoff}_{\text{Greenfield}}$ = $PR_{\text{Greenfield}} \times A \times RD$

= 3940 m^3

Model run with 1 in 100-year rainfall event. Catchment duration T_p = 16 h. vi

$\text{Runoff}_{\text{Development}}$ = 3260 m^3 (model output) xviii

$V_{\text{additional runoff}}$ = 3260 - 3940

= -680 m^3 vii

Because greenfield runoff is (theoretically) more than the runoff from the development, the system does not require any long-term storage. viii

Comments

1. Although the mathematical basis for the analysis is correct, judgement is needed as to whether the rate of runoff is such that there should be some "long-term" storage utilised (see Section 10.5 for alternative analysis options). In addition, it must be recognised that the PR equations (urban and rural) have their limitations in terms of accuracy, even though they represent current best practice.

Notes

v Model run without storage.

vi Storage required in order to reproduce greenfield runoff.

vii This reflects the possible limited accuracy of the two equations used for determining runoff volumes.

viii Model run with five- and 20-year storage to find spill volume for storing additional greenfield runoff.

5. SITE FLOOD PROTECTION (HINCKLEY)

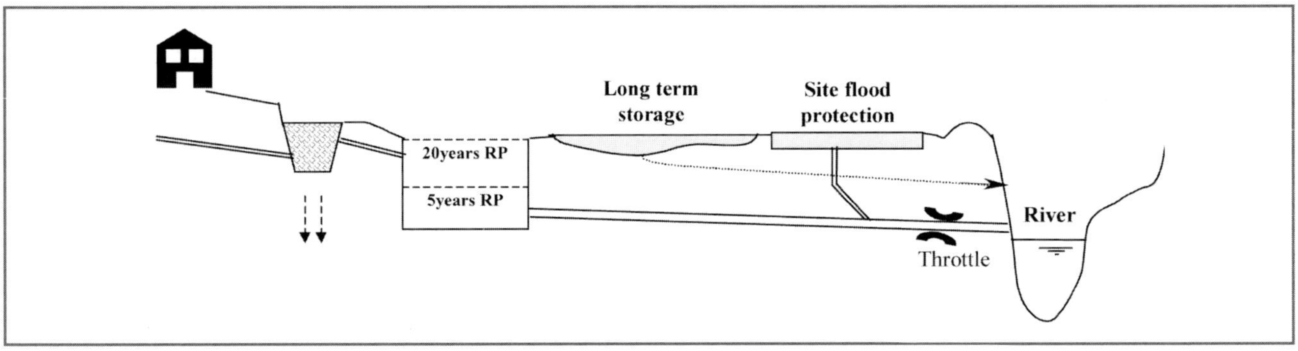

DRAINAGE MODEL

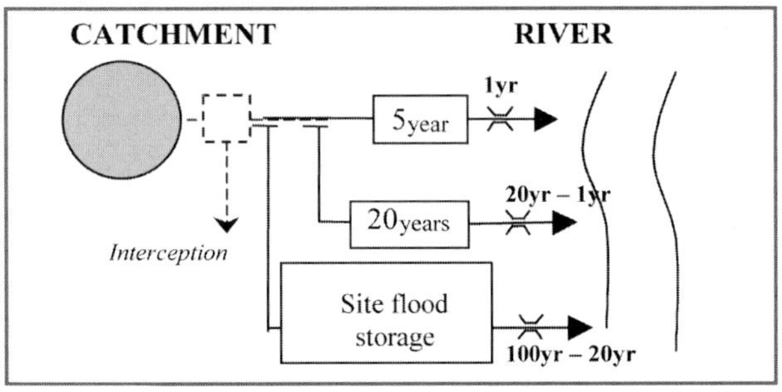

Site protection requires:

Limit discharge = **100**-year greenfield runoff (minimum **10** l/s)

Using IH Report 124 (F = 2.45):

Hinckley limit discharge = 0.122 m³/s

Model run with 1 in 100-year rainfall event. The site-critical duration is four hours.

Summarising:

$V_{\text{interception}}$ = 120 m^3

$V_{\text{5yr storage}}$ = 420 m^3

$V_{\text{20yr storage}}$ = 300 m^3

$V_{\text{100yr long term storage}}$ = 0 m^3

$V_{\text{100yr flood storage}}$ = 470 m^3

As a consequence:

$V_{\text{interception + 5yr+20yr+ 100yr temp.storage}}$ = 1310 m^3

Comments

Notes

CONCLUSION – HINCKLEY

Drainage model

The volume of runoff for greenfield and developed site is virtually the same and therefore (theoretically) there is no long-term storage requirement needed. The results obtained are summarised in the graph below.

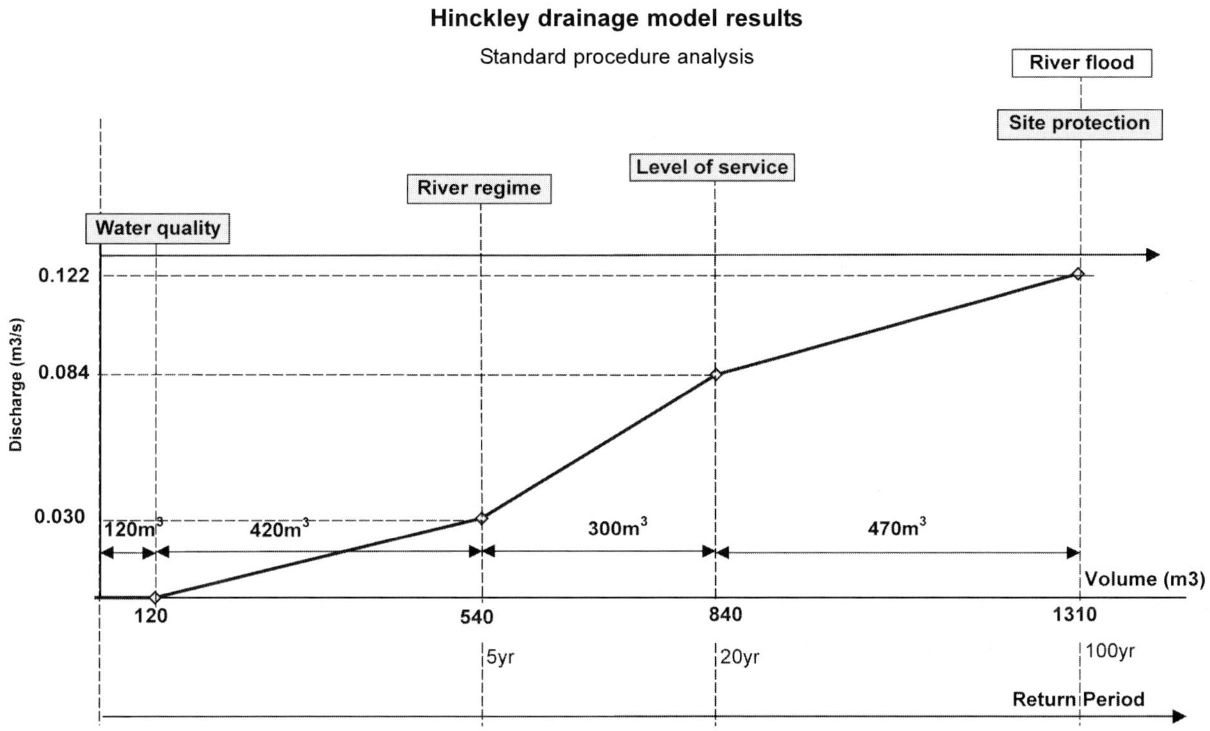

Discussion on an alternative and more conservative assumption on runoff volumes is given in Section 10.5.

TSR analysis

The model was run with the Hinckley 36 years time series rainfall. The results showed that the 20-year storage structure was filled once. On this occasion (rainfall event dated 10 July 1968), 725 m^3 spilled into and passed through the 20-year storage structure. There was no predicted need to use site flood protection storage.

Unfortunately, no flow records are available for the River Sowe before 1978 and therefore it was not possible to check whether the event mobilising the 20-year storage volume occurred during a period of high flow in the river.

These results indicate an under-utilisation of the storage provided.

E THE WALLINGFORD PROCEDURE AND SIMULATION MODELLING

E1 Historical context

The phrase "The Wallingford Procedure" – is regularly encountered by those seeking to get their proposed drainage systems consented. The following explanation describes the original derivation and meaning of the Wallingford Procedure and what it is generally accepted to mean today.

In 1981, HR Wallingford, with assistance from the Institute of Hydrology, completed a DoE-funded project by producing a document of five volumes and a range of software called the WASSP suite of programs. This was called "The Wallingford Procedure".

The programs included the Modified Rational Method to design new sewers, an alternative design method referred to as the Hydrograph Method, and a simulation tool to predict the performance of existing systems. The Modified Rational Method is basically an extension of the Lloyd Davies approach (usually referred to as the Rational Method).

The Hydrograph Method is effectively a computer program for the Hydrograph Method as described by Road Note 35 (TRRL, 1976) in terms of pipe flow routeing, but taking a more sophisticated approach to surface runoff. In addition, it used a statistical runoff module to determine the percentage of rainfall-runoff (usually referred to as the PR equation).

The Simulation Module incorporates the same approach to surface runoff as the Hydrograph Method, but it also explicitly models the effect of manhole storage and the throttling caused by under-sized pipes.

Thus when authorities ask for the Wallingford Procedure to be applied, a range of options is available and it is not always clear whether it is the Modified Rational Method, the Simulation Method, or the runoff model that is being asked for. The tool WASSP has long been superseded, along with some of the original runoff routeing equations. However, current software is effectively applying the same technique to network design and analysis. Details of each of the methods are given in the following sections.

E2 The Modified Rational Method

The Modified Rational Method takes the standard formula:

$$Q = CiA$$

or

$$Q = 2.78CiA \qquad \text{(E1)}$$

when applied using SI units

where:

Q	is in l/s
C	is a unitless coefficient
i	is in mm/h
A	is in ha

but modifies the value C into two coefficients of C_v and C_r.

E2.1 Coefficients C_v and C_r

C_v is referred to as a volumetric coefficient. The volume of runoff is now generally accepted as not being 100 per cent from paved catchments. The value of C_v proposed is the value determined by the PR equation, which is described more fully later. It is now considered that if the value of C_v is only applied to paved surfaces then the coefficient ranges from 0.6 to 0.9 depending on the soil type.

C_r is a routeing coefficient. Theoretical values between 1 and 2 can be determined from time-area diagrams and the peakedness of the rainfall depending on the catchment shape. It was decided that a fixed value of 1.3 should be used for all catchments.

Advice on the time of entry (TE) is given. It states that TE reduces with increasing return period and steeper catchments. Ranges for TE are given as 3–10 minutes. A figure of 5 minutes would generally be acceptable for most design situations.

E2.2 Rainfall intensity

Information on the intensity-duration-frequency curves of rainfall for use in applying the Rational Method is often given in the form:

$$D = at/(b+t) \qquad \text{(E2)}$$

or

$$D = at/(b+t^n) \qquad \text{(E3)}$$

where:

D	is rainfall depth
t	is time
a,b,n	are constants.

This is referred to as the Talbot formula. Information on rainfall at any location in the world is usually available in this form.

In the UK, the work that resulted in the *Flood studies report* (IH, 1975) gave a more detailed method of deriving this information for any point in the UK using key rainfall characteristics for any location. These parameters are:

M_560	The five-year return period depth of rainfall for 60 minutes
r	The ratio of the M_560 minute and M_52 day depths

These two values can be obtained from figures given in Volume 4 or the maps in Volume 5 of the Wallingford Procedure.

From these values, look-up tables in Volume 4 can be used to derive rainfall intensities for any return period and any time. In addition, an areal reduction factor (ARF) table is available to reduce the intensity to take account of the spatial reduction of the average rainfall intensity.

In 1999 the FSR was superseded by the *Flood estimation handbook* (CEH, 1999), which was effectively a repeat of the FSR research with more information and new statistical analytic procedures.

In practice, the industry often uses a fixed figure of 35 mm/h or 50 mm/h for designing the size of sewers. This is because the time period for intensities, which are greater than this value, are short and therefore have little practical effect. This results in conservative pipe sizes and gradients, but, due to the size of small developments, applying this approach has little financial impact.

E3 The Hydrograph Method

The principal difference between Rational Method and Hydrograph Method is the use of rainfall in the form of hyetographs in place of intensity-duration-frequency (IDF) data. This means that the method is a volumetric method rather than one based on peak flow.

The Hydrograph Method in the Wallingford Procedure for design of pipes applies the surface runoff model and percentage runoff model described later in the Simulation Method (see Section E4). Because it is a design method it takes no account of existing pipes in the system. It also has a very simplistic model for dealing with overflows.

E4 Simulation modelling analysis

The Simulation Method uses rainfall in the form of a hyetograph, which is processed for runoff by the statistical runoff equation (PR), which is routed on the surface and then routed through the pipe system. Because the method is volumetric, storage within networks is explicitly catered for and the effects of throttles determined. This section describes this method and discusses its limitations and advantages. This will assist developers and their technical advisers to understand what is being asked for and to discuss drainage design requirements with the authorities.

The generation of runoff and therefore pipe sizes (and storage tank or pond sizes) are directly a function of runoff. Therefore, the runoff aspect of the Wallingford Procedure is covered in some depth to allow an understanding of this important aspect to be obtained.

E.4.1 Rainfall

The rainfall hyetographs used are those based on the work carried out by the Flood Studies exercise. The hyetographs are symmetrical bell-shaped curves where the total depth and peakedness of the rainfall are functions of return period and duration of the event. In addition, differentiation is made between summer and winter profiles. Figure E1 illustrates summer and winter rainfall profiles and also the effect of different durations of design events.

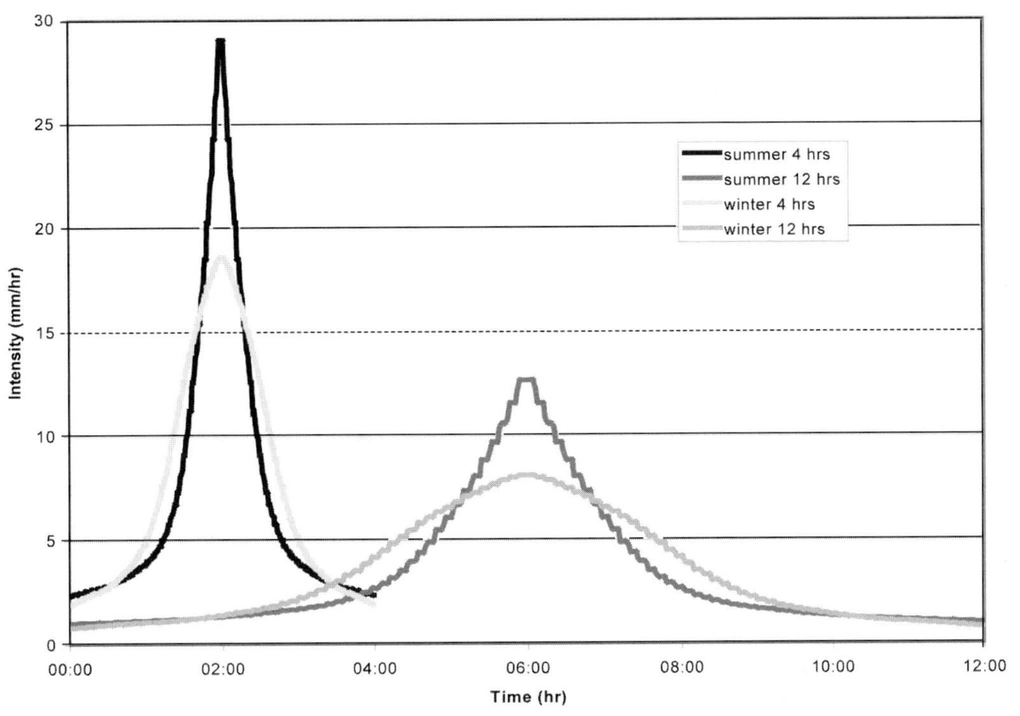

Figure E1 *Five-year summer and winter rainfall hyetographs – four-hour and 12-hour durations*

Rainfall characteristics vary across the country and the storm profile is obtained from two parameters that can be obtained from the rainfall maps given in Volume 5 of the Wallingford Procedure. These parameters are:

$M_5 60$, which is the five-year 60-minute rainfall depth

Rainfall ratio "r", which is the ratio of the five-year, 60-minute depth and the five-year, two-day depth of rain.

Figures E2 and E3 illustrate values found across the country for these two parameters. More information on the derivation of the rainfall profiles can be obtained from Volume 1 of the Wallingford Procedure. The detailed maps are available as Volume 5 of the *Wallingford Procedure* (HR Wallingford and IH, 1981a), or as part of the newer *Wallingford Procedure for Europe* (Kellagher, 2000).

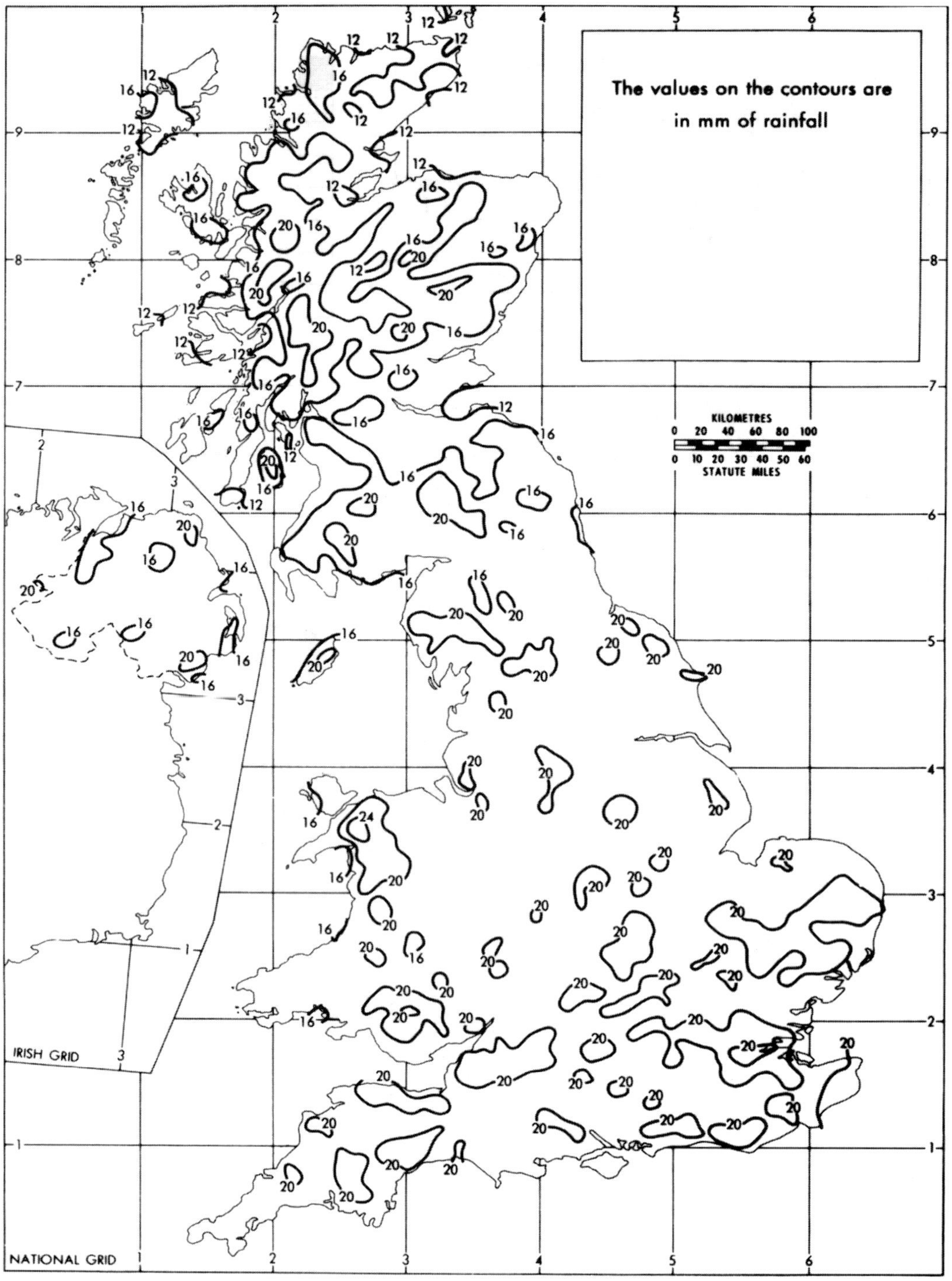

Figure E2 *Rainfall depths of five-year return period and 60 minutes duration ($M_5 60$ min)*

In assessing pipe networks, new or existing, summer events are generally used. When storage is being assessed, winter profiles are used. More discussion on this aspect is given in Chapter 10.

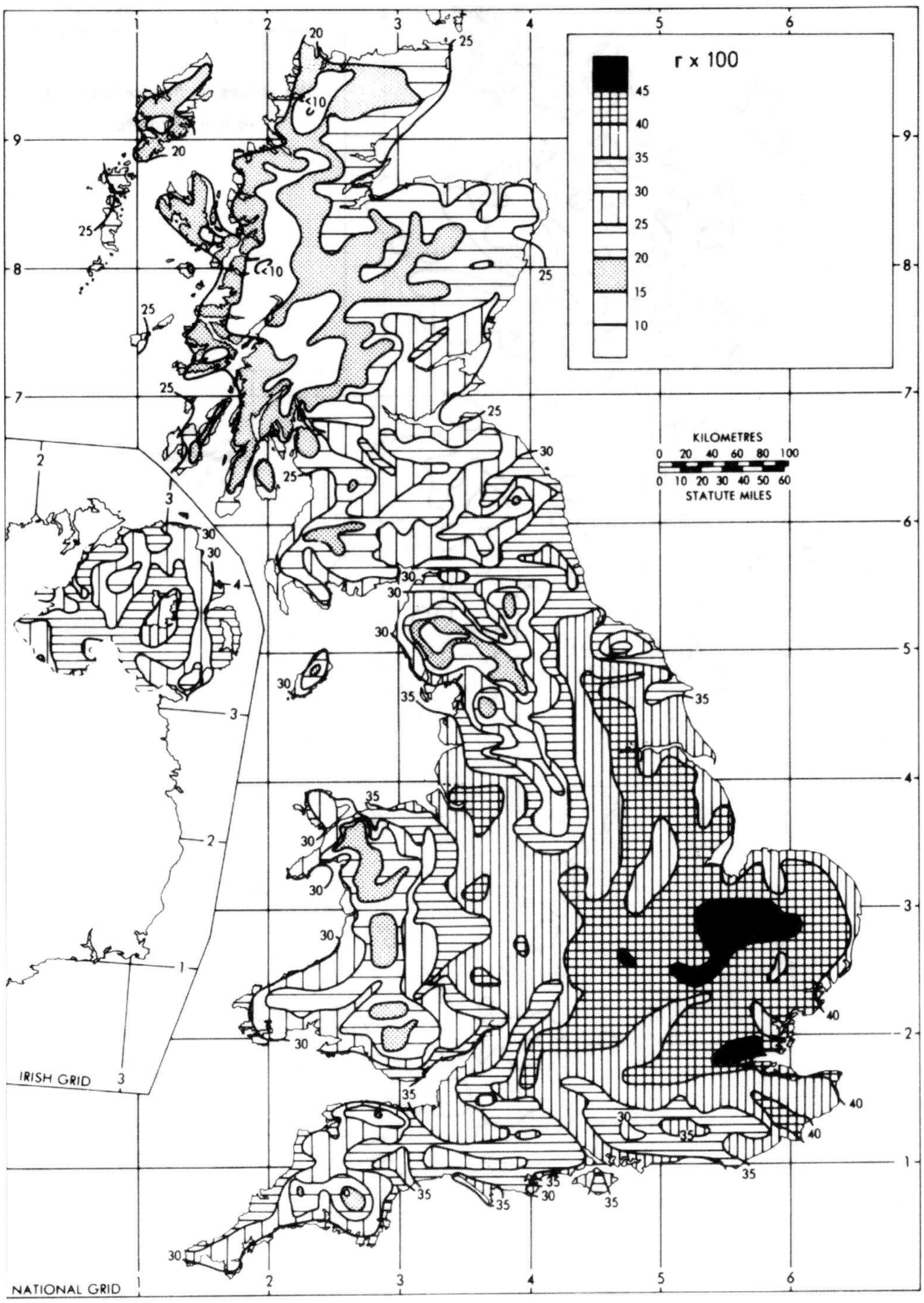

Figure E3 *Ratio of 60-minute to two-day rainfalls of five-year return period (available from the Wallingford Procedure)*

E4.2 Loss models

Before going into the detail of the PR runoff models, it is important to make a brief reference to the fact that not all rainfall is made to runoff. In addition to the proportion that is "lost" by infiltration or evaporation, there is the concept of depression storage, which has to fill up before runoff takes place. This depression storage is modelled and is a function of catchment slope. It is defined as:

$$D = k/(s)^{0.5} \tag{E4}$$

where:

D	is in mm
k	is a constant
s	is catchment slope.

Typical values for depression storage are in the region of 0.5–2 mm.

The relevance of this feature is not very great when using design events, as the depression storage is small compared with the depth of rainfall.

E.4.3 The PR equation

The Old fixed PR Wallingford Procedure equation has been modified twice since it was brought out in 1981.

The Old PR equation is the standard runoff model used to represent continuing losses for UK urban catchments and is applied with the initial losses model described previously. Runoff losses are assumed to be constant throughout a rainfall event (runoff does not increase as the catchment gets wetter) and are defined by the relationship:

$$PR = 0.829\,PIMP + 25.0\,SOIL + 0.078\,UCWI - 20.7 \tag{E5}$$

where:

PR	= percentage runoff
PIMP	= percentage impermeability
SOIL	= an index of the water holding capacity of the soil
UCWI	= Urban Catchment Wetness Index.

The PR equation was derived by statistical analysis from data from 33 catchments. It should be noted that the equation is entirely statistical and takes no account of ground contouring.

An explanation of the meaning and derivation of these parameters follows.

PIMP

This parameter is the percentage imperviousness of the catchment obtained by dividing the total directly connected impervious area (both roofs and roads) by the total contributing area.

SOIL

The soil index SOIL is based on the winter rain acceptance parameter (WRAP) included in the *Flood studies report* and can be obtained from the *FSR* (IH, 1975), *The Wallingford Procedure* (HR Wallingford and IH, 1981a), or *The Wallingford Procedure for Europe* (Kellagher, 2000). Figure E4 shows a copy of the map. The index broadly describes infiltration potential and was derived by a consideration of soil permeability, topographic slope and the likelihood of impermeable layers. The five classes of soil recognised are shown in Table E1 and Figure E4.

Table E1 *Different classes of soil*

Soil class	WRAP	Runoff	SOIL	Soil type
1	Very high	Very low	0.15	Sandy, well-drained
2	High	Low	0.30	Intermediate soils (sandy)
3	Moderate	Moderate	0.40	Intermediate soils (silty)
4	Low	High	0.45	Clayey, poorly drained
5	Very low	Very high	0.50	Steep, rocky areas

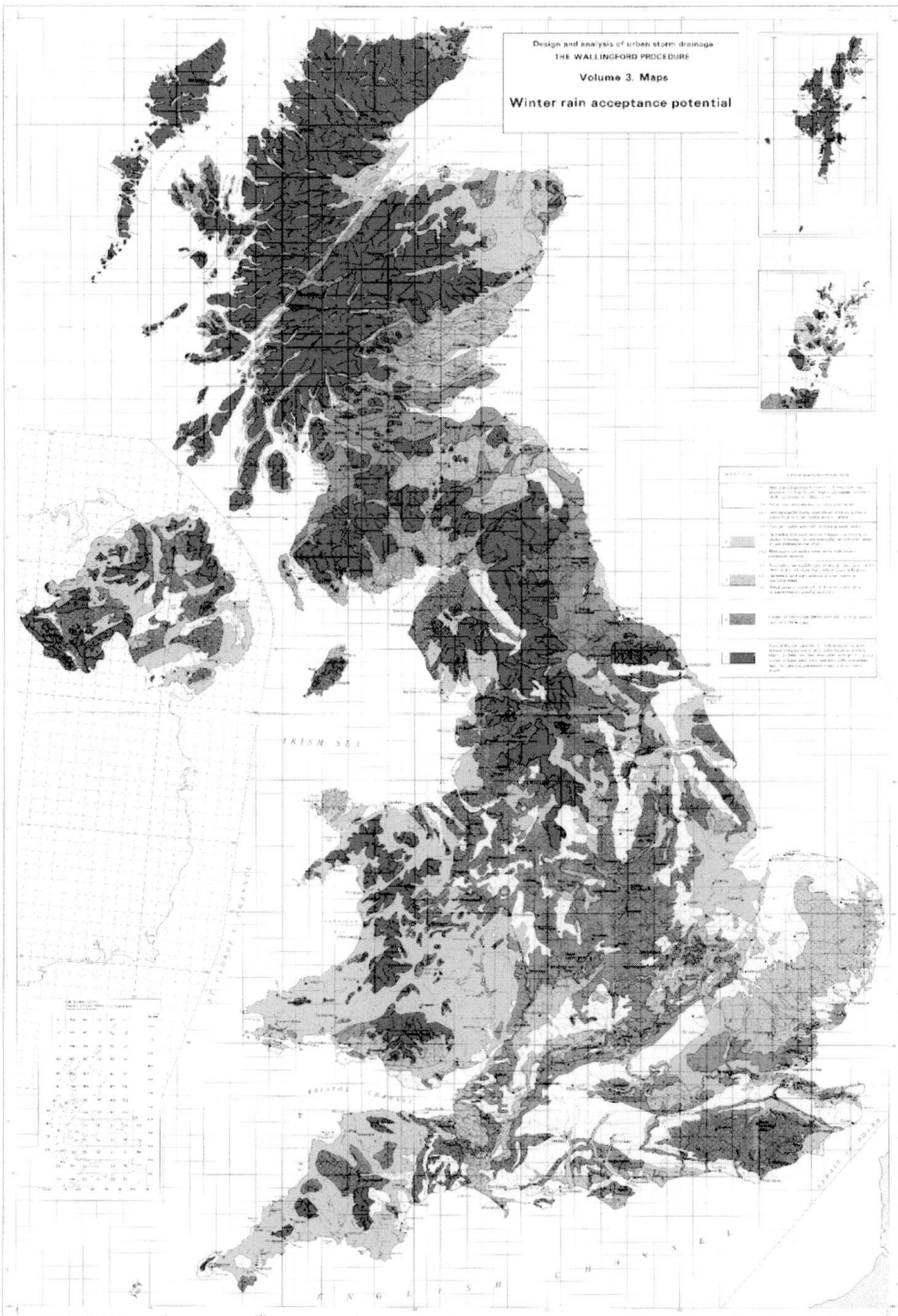

Figure E.4 *WRAP map of UK (available from the Wallingford Procedure)*

UCWI

This is the Urban Catchment Wetness Index, which is a composite of two antecedent wetness parameters and is given by:

$$UCWI = 125 + 8API_5 - SMD \tag{E6}$$

where:

API_5 = five-day antecedent precipitation index (mm)

SMD = soil moisture deficit.

The value for UCWI is calculated from these parameters for specific events, but design values are provided by referring to a figure relating UCWI to the annual average rainfall for that location (Figure E5). Values are provided for both winter and summer conditions. Figure E6 shows the annual average rainfall found in UK.

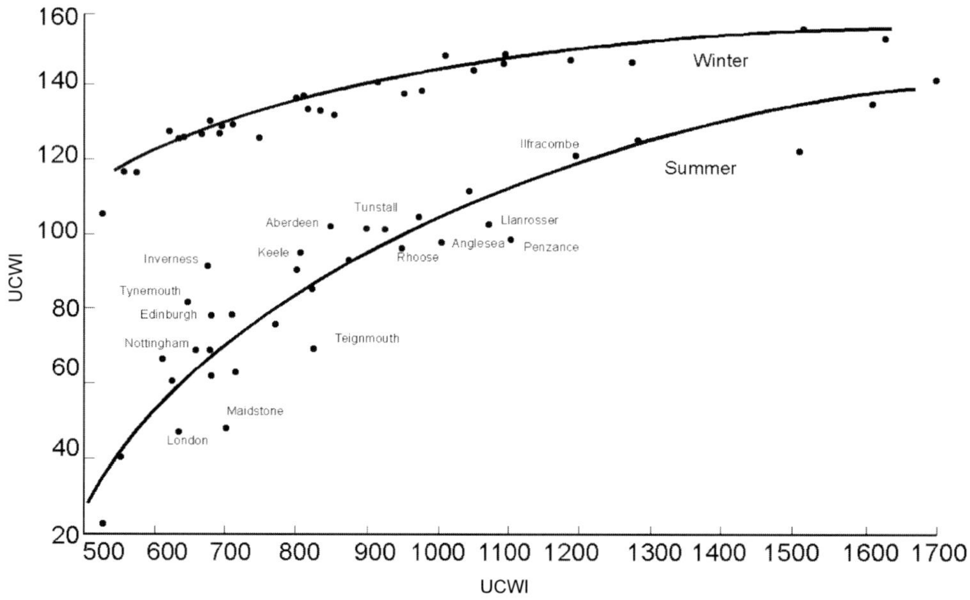

Figure E5 *Seasonal UCWI relationship with SAAR*

For specific events, API_5 is calculated using the following procedure. First determine the rainfall depths (in mm) for the five days prior to the event. The API_5 value at 09.00 on the day of the event is then defined by:

$$API5_9 = \Sigma_{n=1,5} P_{-n} C_p^{n-0.5} \tag{E7}$$

where:

P_{-n} = rainfall on day n before the event

C_p = decay coefficient = 0.5.

Finally the API_5 at the time of the event is given by:

$$API_5 = API5_9 C_p^{(t'-9)/24} + P_{t'-9} C_p^{(t'-9)/48} \tag{E8}$$

where:

t' = time (hours) of the beginning of the event

$P_{t'-9}$ = rainfall depth between time t' and 09.00.

The soil moisture deficit is calculated from a similar equation:

$$SMD = SMD_9 - P_{t'-9} \qquad (E9)$$

where:

SMD_9 = soil moisture deficit at 09.00 on the day of the event.

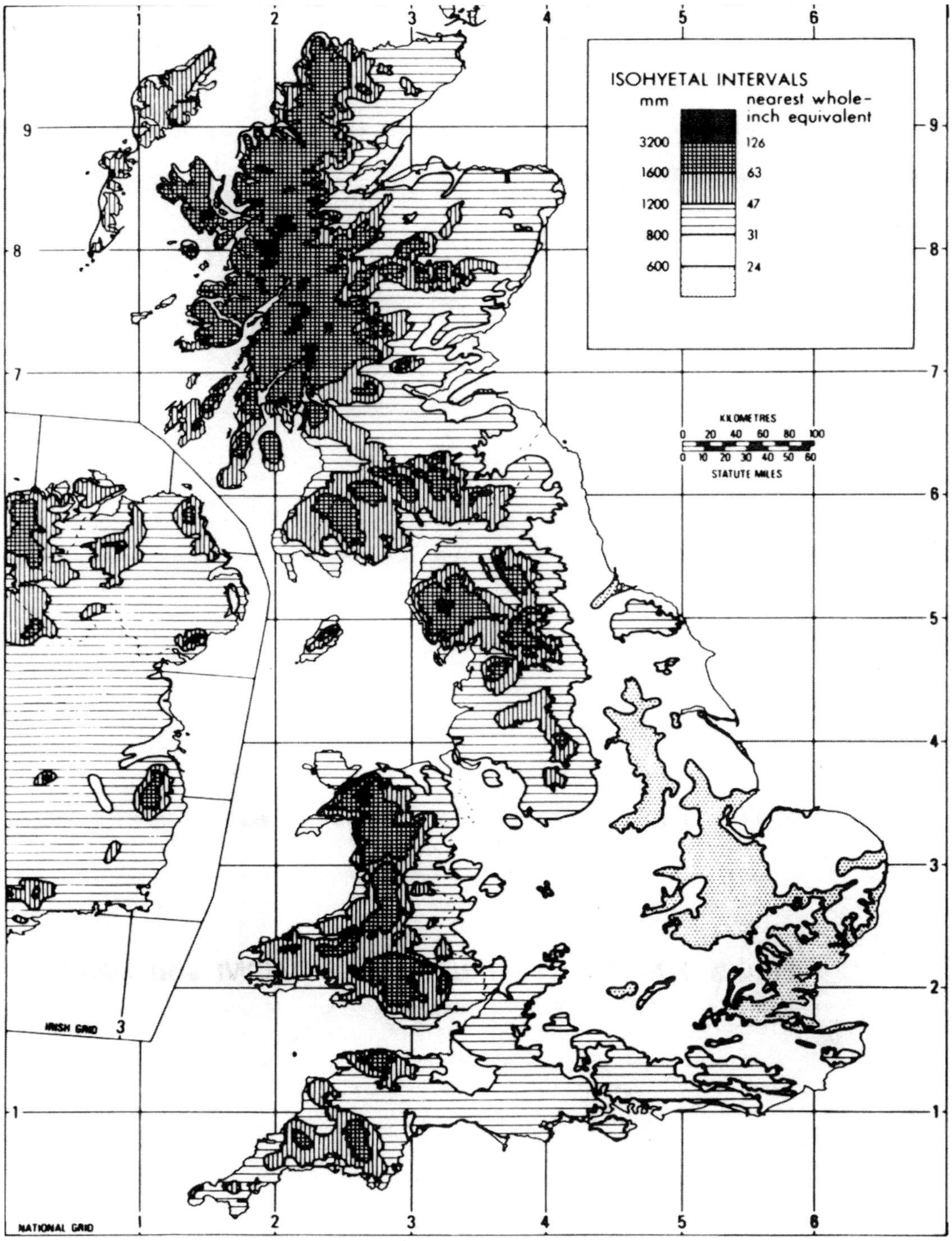

Figure E6 *Annual average rainfall in UK (available from the Wallingford Procedure)*

The SMD_9 value (known as ESMD) was obtainable from the UK Meteorological Office until 1997. It was calculated from a water balance between daily rainfall and an estimate of evapotranspiration based on the use of Penman's equation, assuming a notional catchment under short-rooted vegetation (50 per cent), long-rooted vegetation (30 per cent) and riparian areas (20 per cent). Since the development of the procedure, the Meteorological Office has ceased the routine issue of ESMD and issues a new SMD value based upon the use of a different calculating system (MORECS), which is a modification of the Penman equation by Monteith.

In general, there appears to be little practical difference between the use of the two values, with little consistent bias.

Inspection of the Old fixed PR Wallingford Procedure equation indicates that for low values of PIMP, SOIL and UCWI, low or even negative values of PR can be predicted. Consequently, a minimum value of PR_{paved} of 20 per cent together with a maximum of 100 per cent is specified. It should be appreciated that unrealistic PR values can be predicted with low values of SOIL (eg 0.15) in combination with both low values of PIMP (eg PIMP < 30 per cent) and UCWI. Its application on sewers with partially separated systems is generally inappropriate, therefore. Figure E7 illustrates how PR changes with PIMP and SOIL.

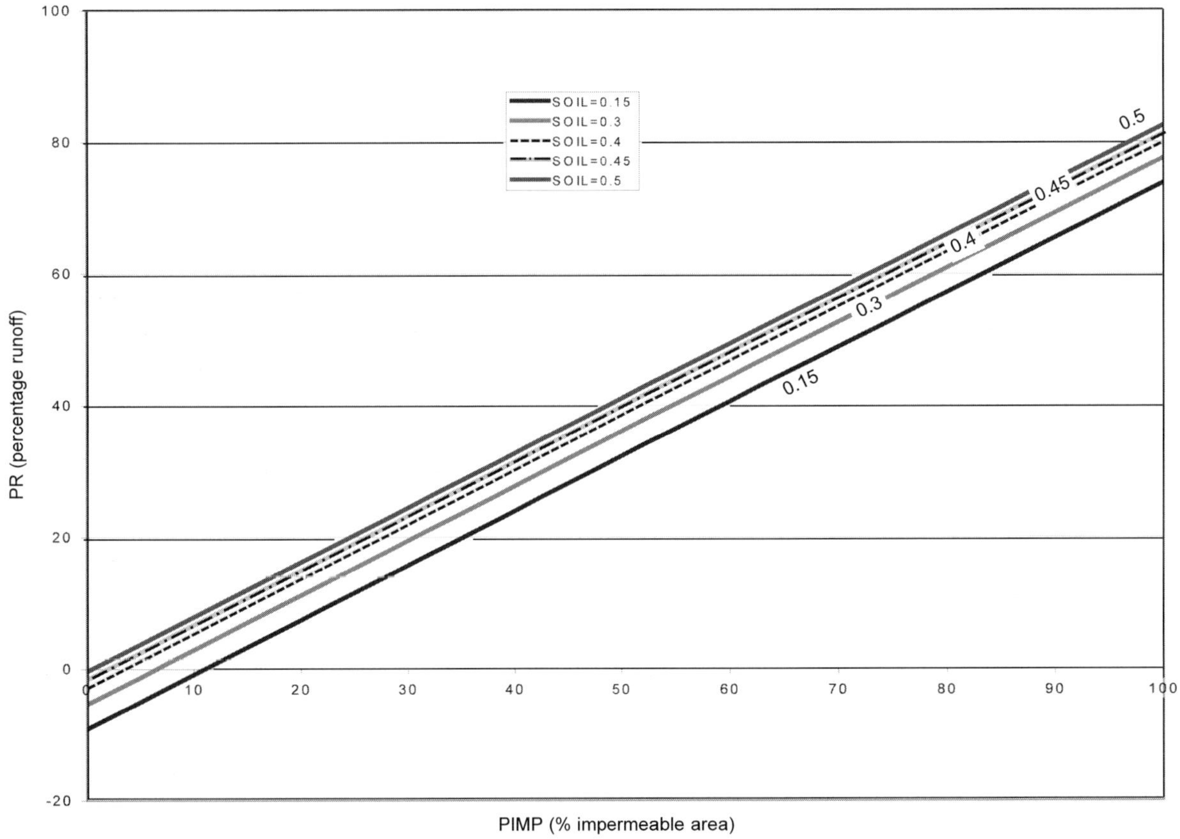

Figure E7 *PR as a function of SOIL and PIMP (Old fixed PR Wallingford Procedure equation)*

It should be noted here that the Old fixed PR Wallingford Procedure equation is still the most popular PR equation. Volume 1 of *The Wallingford Procedure* (HR Wallingford and IH, 1981a) refers to PR being a minimum of 0.4 PIMP and this is now not applied. The current rule is that PR_{paved} is a minimum value of 20 per cent to avoid totally inappropriate runoff values when PIMP is low. Similarly, the distribution of runoff to paved and pervious surfaces has changed from that described in Volume 1 and is presented here.

Runoff distribution for different surfaces

This model predicts the total runoff from all surfaces in the sub-catchment, including both pervious and impervious. Runoff for the entire catchment is distributed between the different surfaces using weighting coefficients. All surfaces can therefore contribute some runoff even at low runoff rates, provided that initial losses have been satisfied.

The weighting is carried out as follows:

$$PR_i = \frac{f_i A_i}{\Sigma_{n=1,3} f_n A_n} .PR \qquad (E10)$$

where:

f_i = weighting coefficient for surface i

PR_i = percentage runoff for surface i

A_i = area for surface i.

Default parameters for the weighting coefficients are shown in Table E2. These values can be changed but are rarely altered.

Table E2 *Typical weighting coefficients*

Weighting coefficient	Surface	Value
F_1	Paved	1.0
F_2	Roofed	1.0
F_3	Pervious	0.1

E4.4 The new PR equation

The new UK PR equation was developed jointly by HR Wallingford, the Water Research Centre and the Institute of Hydrology with support from North West Water PLC. It has been designed to replace the familiar Old fixed PR Wallingford Procedure equation defined previously. It is becoming more widely used, although the old equation remains more popular.

The New variable PR Wallingford Procedure equation was designed primarily to overcome some of the difficulties experienced in practical application of the old equation, namely:

- the Old fixed PR Wallingford Procedure equation defines PR as being a constant throughout a rainfall event irrespective of catchment wetness. Clearly. for long-duration storms, lower losses towards the end of the event may be significant in terms of urban drainage design
- problems have been encountered in applying the PR equation to partially separate catchments and to catchments with low PIMP and low SOIL values.

To overcome these problems various new model forms were investigated using a subset of the original data.

The advised model derived by this analysis is of the form:

$$PR = IF*PIMP + (100 - IF*PIMP)*\frac{NAPI}{PF} \qquad (E11)$$

where:

IF = effective impervious area factor

PF = moisture depth parameter (mm)

NAPI = 30-day antecedent precipitation index

This equation divides PR into two elements. First, the impervious area runoff is obtained by using an effective contributing area factor, IF. After initial losses on impervious surfaces, remaining losses are therefore given as a constant fraction of rainfall volume.

Recommended values of IF are indicated in Table E3 and can be compared with the PR_{imp} values for the individual catchments derived using the old PR equation. One of the principal features of this equation (and a possible drawback) is that engineers have to choose a value.

Table E3 *Recommended values of IF*

Surface condition	Effective impervious area factor, IF
Poor	0.45
Fair	0.60
Good	0.75

The losses on pervious surfaces and also non-effective impervious areas are represented by the second term of the equation. The first part of this term represents the total percentage of the catchment occupied by pervious and non-effective impervious areas. The losses from this area are dependent on the function NAPI/PF. NAPI is defined as a 30-day API with evapotranspiration and initial losses subtracted from rainfall. As for API_5, API_{30} is given by:

$$API_{30} = \sum_{n=1,30} P_{-n}\, C_p^{n-0.5} \tag{E12}$$

The constant value C of the API has been made dependent on the soil type to reflect the faster reduction of soil moisture on lighter soils. The relationship between C and soil type is shown in Table E4.

Table E4 *Relationship between SOIL type and decay coefficient "C"*

SOIL type	C
1	0.1
2	0.5
3	0.7
4	0.9
5	0.99

The moisture depth parameter, PF, was calibrated using the data described above. A value of 200 mm was obtained (which compares well with the available water capacity of soils with grass vegetation). It is dangerous to modify this value without careful consideration of the consequences.

Figure E8 illustrates the effect of increasing rainfall on percentage runoff using the New variable PR Wallingford Procedure equation. This should be compared to Figure E7. The assumptions used in the figure are as follows:

Old PR

PIMP	= 50 per cent
SOIL	= 1–5
UCWI	= 100
Rainfall	= 50-year 18-hour summer event (78 mm)

New PR

SOIL	= 1–5
NAPI	= 0 mm at start of the event
PF	= 200 mm
IF	= 70 per cent

Rainfall

M_560	= 20 mm
Rainfall ratio "r"	= 0.4

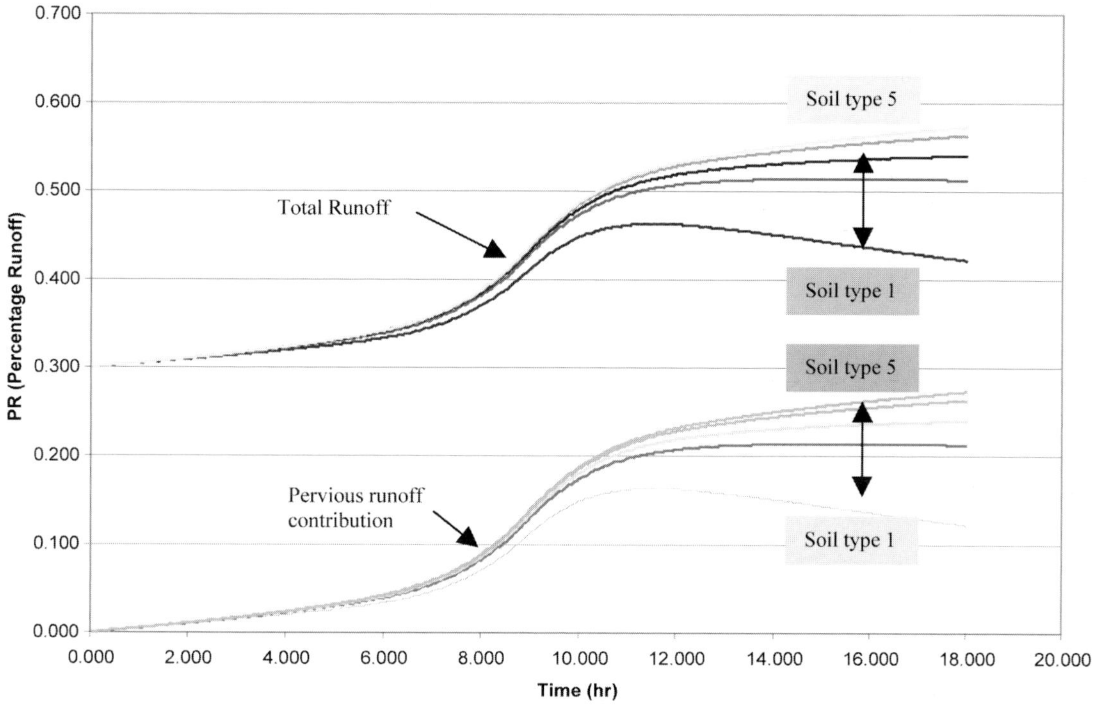

Figure E8 *Percentage runoff as a function of rainfall depth using the New variable PR Wallingford Procedure equation*

E4.5 Surface routeing

The rainfall runoff is not passed directly into the pipe model, but first goes through an attenuation process using a double linear reservoir. Again, this modifies Volume 1 of the procedure where a non-linear reservoir is used.

The model effectively passes the runoff through two conceptual reservoirs on the surface before passing the flow into a node of the system (a manhole).

The routeing of this flow is a function of:

- catchment area
- catchment slope
- rainfall intensity
- a coefficient for surface type.

These parameters allow the effects of different catchment characteristics to be represented to alter the flow rate into the sewer for each surface type.

E4.6 Pipe flow

The original pipe routeing model used the Muskingum Cunge method. This was fast and stable, but had limitations when it considered pipes with flatter gradients or reverse flows at overflows.

Modern software uses forms of solving the St Venant full solution equation. This is achieved by applying a conceptual slot, referred to as the Preissman slot, in the sewer to allow the solution always to consider the flow to have a free surface. There are different solution techniques that can be used to solve the equation, but mathematical precautions need to be taken to ensure stability.

All sewer models take into account pipe roughness, headloss at structures and bends so that all aspects of pipe flow can be accurately represented.

F THE COLEBROOK-WHITE EQUATION FOR THE DESIGN OF SEWERS

F1 Introduction

1. There are several velocity equations commonly used to design pipes. These are:

   (a) Manning: v = $\frac{1}{n} R^{2/3} s^{1/2}$ (metric units)

   (b) Hazen-Williams: v = C $R^{0.63}$ $s^{0.54}$ (metric units)

   (c) Darcy-Weisbach: v = $\left(\frac{8gRs}{\lambda}\right)^{1/2}$

   where:

v	= mean velocity m/s
R	= hydraulic radius (= D/4 for circular pipes flowing full), m
D	= pipe diameter, m
s	= energy gradient
n, C	= are dimensional friction coefficients
λ	= is a dimensionless friction coefficient, often known as the friction factor.

2. All these equations are simple to use. The principal difficulty lies in the choice of a suitable friction coefficient. Manning and Hazen-Williams originally developed their empirical equations from relatively narrow ranges of flow conditions, in which coefficients n and C were functions only of the roughness of the surface of the pipe or channel. In practice, however, these equations are used over a much wider range, and experiments have shown that the coefficients in all three equations listed above are functions of pipe roughness.

F2 Colebrook-White equation

3. Colebrook, in collaboration with White (Colebrook, 1939), examined experimental data for flow in commercial pipes. Relating these to theoretical work on turbulent flow they produced the following equations:

$$\frac{1}{\sqrt{\lambda}} = -2\log_{10}\left[\frac{k}{14.8R} + \frac{2.51v}{R\sqrt{128gRs}}\right]$$

   where k is a roughness height and v is the kinematic viscosity of the fluid. Although the roughness value, k, has the dimensions of length it cannot in general be directly related to physical measurements of the pipe surface and hydraulic experiments must be carried out to determine its value.

   Later experiments described by Acker (1958) and Lamont (1954) confirmed that the equation accurately describes the flow relationships for a wide range of channels and pipes, including all the data on which the Manning and Hazen-Williams equations were based. However, the equation is not suitable for corrugated pipes or for pipes and channels containing sediment deposits. For the past 20 years the Colebrook-White equation has been accepted as the most reliable and accurate one that is available.

F3 Exponential equations

4. The main objection to the Colebrook-White equation is that it is not easy to handle, although the strength of this objection should have less validity now that calculators are common. Ackers has shown that it can be replaced by an exponential equation of the form $v = C\ R^{\alpha} s^{\beta}$ where α and β are functions of R/k and vk/ν.

A given exponential equation will only be exactly equivalent to the Colebrook-White equation at one particular combination of these parameter values, but it will be accurate within a few per cent if R or v do not depart by more than four-fold from the ideal value. If the exponential equation is applied well outside its limits of application, it is necessary to manipulate the value of coefficient C' so that it will give the correct answer.

F4 Manning equation

5. Ackers has shown that the Manning equation is a good approximation to the Colebrook-White equation when the relative roughness R/k lies between the limits of 7 and 130. The appropriate Manning n for a particular pipe material should be calculated from $n = k^{1/16}/83.3$ (k in mm). A further restriction on the use of the Manning equation is that the flow should be in the fully turbulent flow region ie vk/ν > 807. The following table, which assumes water at 15°C at which ν is $1.143 \times 10^{-6} m^2/s$, illustrates what this means in practical terms.

Description	Roughness k (mm)	Applicable range of diameters (mm)	Velocity should exceed (m/s)
Coated steel, clayware	0.06	1.7–31	15
Concrete	0.15	4.2–78	6
Slimed sewer. Water main with slight tuberculation	1.5	42–780	0.6
Slimed sewer. Water main with moderate tuberculation	3.0	84–1560	0.3

The conclusion is that the Manning equation has only a very restricted application to the flow in pipes and sewers. The Crimp and Bruges equation is even more restricted in application. It corresponds to the Manning equation with an n value of 0.012, or a k value of 1 mm.

F5 Hazen-Williams equation

6. The Hazen-Williams equation is equivalent to the Colebrook-White equation at a vk/ν value of 11.2 and a relative roughness, R/k, of 4.1×10^4. At a velocity of 1 m/s, these values correspond to a 2.18 m-diameter pipe with a roughness value of k = 0.013 mm (C = 130 in metric units), eg asbestos cement in good condition; at 2 m/s, to a diameter of 1.09 m and roughness value of k = 0.007 mm (C = 130), eg smooth pipes such as glass, perspex, brass; at 3 m/s to a diameter of 0.73 m and roughness value of k = 0.005 mm (C = 131) for a very smooth pipe. To use the Hazen-Williams equation outside this restricted range, ie to apply it to the normal commercial pipes used by the drainage and water supply industry, it is necessary to use tables. These give values of C as a function of pipe material, diameter and velocity, but are only approximate for the rougher classes of water main. A further complication is that the Hazen-Williams equation cannot be used to determine the diameter directly, given the discharge and the slope. The friction coefficient must first be assumed, the diameter calculated and the coefficient then adjusted to correspond to the appropriate coefficient for the computed diameter. This procedure continues by successive approximations until the adjusted coefficient agrees with the coefficient appropriate to the calculated pipe diameter.

F6 Use of the Colebrook-White equation

7. The great advantage of the Colebrook-White equation is that it is consistent with the theory of turbulence and with experiments on pipe friction. The roughness height k is independent of the flow conditions and is a function only of the pipe surface condition. One objection to the equation is that it can only be used to determine v, given k, R (or D) and s. A successive approximation solution must be used if "D" or "s" has to be calculated. However, there are explicit equations that are equivalent to the Colebrook-White equation and these allow any variable to be determined, given the other three variables. These explicit equations are as follows:

   (a) To find v, given k, D and s

$$v = -\sqrt{32gRs}\,\log_{10}\left[\frac{k}{14.8R} + \frac{1.255v}{R\sqrt{32gRs}}\right]$$

   (b) To find s, given k, D and v

$$s = \frac{v^2}{32gR}\,\frac{1}{\left[\log_{10}\left|\frac{k}{14.8R} + \frac{5.13v^{0.89}}{(4vR)^{0.89}}\right|\right]^2}$$

   (c) To find D, given k, v and s

$$\frac{D}{4} = \frac{v^2}{32gs}\,\frac{1}{\left[\log_{10}\left|\frac{1.558}{(v^2/2gsk)^{0.8}} + \frac{15.045}{(v^3/2gsv)^{0.73}}\right|\right]^2}$$

   Although these equations appear complicated, they are easy to solve with a calculator or a computer. If a manual method of solution is required, charts and tables have been published (available from HR Wallingford) that enable a pipe to be designed given any three out of the four variables.

F7 Conclusion

8. Exponential equations have the attraction of ease of use, but any particular exponential equation will apply only over a limited range of flow conditions. On the other hand, the Colebrook-White equation covers the whole range of turbulent flow.

G SUSTAINABLE DRAINAGE SYSTEMS

G1 Water butts

G1.1 Description and purpose

Water butts are usually used to collect rainwater runoff from the roofs of domestic developments. Use of water butts is rarely considered a SUDS technique and there are several arguments against it. Butts are normally quite small, with a capacity of less than 0.5 m^3, and in wet periods (winter) they are often full, resulting in no attenuation or reduction in the outflow. Conversely, butts store insufficient water during dry periods to provide significant benefits for garden watering or other reuse purposes. It is therefore difficult to persuade house-owners of their benefits. However, a simple proposal by Sheffield Hallam University suggests that they could be designed to attenuate runoff during an event by using a throttle into the drainage system, above which there is a storage volume to accommodate the temporarily stored runoff. If every roof was served in this manner there would be a significant attenuation to roof runoff. Figure G1 shows a schematic of an attenuating water butt.

There are some risks to consider, however. The throttle (which by definition must be small) could become blocked. The attenuation storage volume needs to be carefully designed to ensure that there is sufficient capacity to cater for the design event. This would vary depending on the region as well as the size of the roof area being served. There will be no effective attenuation if this storage is filled before the peak of the storm takes place (assuming the water butt does not just overflow to the garden). Finally, as with all private system proposals, there is a risk that this mechanism will be bypassed or modified by the householder. For all these reasons engineers carrying out calculations for sizing downstream collection systems would generally have to assume that the water butt was full when considering the impact of an extreme event.

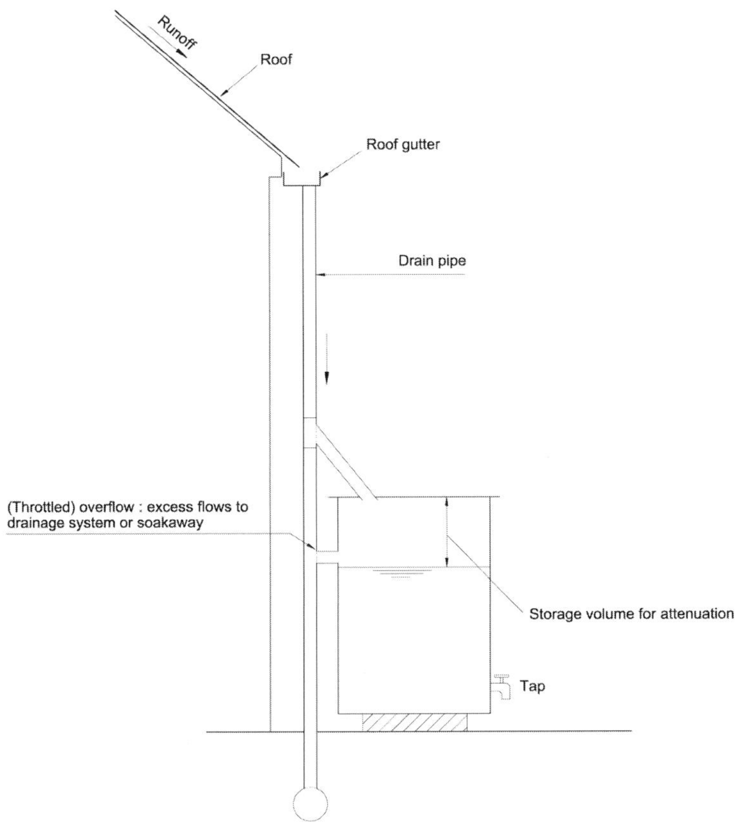

Figure G1 *Water butt*

The other benefit of the water butt is the possibility of reuse. Pollutants in the water are generally low, and several schemes have successfully recycled rainwater for toilet use. This has downstream benefits for reducing stormwater and also reduces water demand.

The measured pollutants from roofs are usually quite low, with the exception of ammonia concentrations derived from the bird droppings and detritus that collect in gutters. Although this limits the benefit of reducing pollutants passing downstream, the low pollutant concentrations make the concept of recycling more attractive, because the water will have fewer odour- and health-related problems even when retained for several days.

G1.2 Design principles

Water butts are designed for two reasons. The first is to maximise the use of water for gardening or recycling for toilet flushing, for example. The second is to attenuate rainfall runoff to protect the system downstream. The size of the excess storage is a function of the inflow/outflow relationship: the inflow is the roof runoff and the outflow is controlled by the throttle size. In principle, the assessment of effective storage needed for either is best achieved using a rainfall time series (at a high temporal resolution), but design storms will provide a fair approximation of the storage requirement for the latter. However, as the dry period is an important issue for determining the volume needed for either the toilet or garden, a design storm approach will not provide sufficient information on the likely system performance.

The basis for garden use or toilet recycling will be to use a very low return period. This is because it would not be cost-effective to use a return period of more than one year, and also the water quality implications of stored water in terms of smell are considerably increased with long retention times.

The criteria used for drainage system design for a site is in the order of 30 years (no flooding). It is unlikely that water butts would be effective in attenuating of flows for these extreme events. It would be unwise to assume a reduction in downstream network capacity requirements on the basis that all houses had water butts unless detailed calculations showed that a reduction in peak flow rate could be achieved. A similar position should be taken for storage design.

CIRIA C521 (Martin *et al*, 2000a) treats water butts as a good housekeeping measure and does not give a methodology for their design and use.

G1.3 Maintenance requirements

Maintenance requirements are minimal and are the responsibility of the householder. To encourage environment-friendly practices, however, local authorities often offer free assistance and occasionally provide and install the unit for a nominal fee.

G2 Sub-pavement storage

G2.1 Description and purpose

Sub-pavement storage has been separated from permeable pavements (Section G11) because this type of storage can be utilised whether the pavement is designed to be permeable or not. Nevertheless, the section on permeable pavement should be referred to for additional information. The concept is that runoff, usually from roofs or a car park, is drained into the sub-base media, which will have been designed to have a high voids ratio. This media has traditionally been formed from single-sized stone, but more recently plastic high-strength, high voids material (90 per cent or greater) has been used. The water is transferred to and from, and sometimes distributed within, the media by a system of perforated pipes. This system results in a significant attenuation of the inflow. In addition to the hydraulic benefits, the water quality properties of the effluent have been found to be considerably enhanced.

Sub-pavement storage facilities often include an impermeable membrane below the pavement and sub-base to prevent infiltration (which is the alternative option where storage is not required). Protection of the groundwater from risk of contamination, particularly where important aquifers exist, usually requires the use of a membrane. Sub-pavement storage is illustrated in Figure G2.

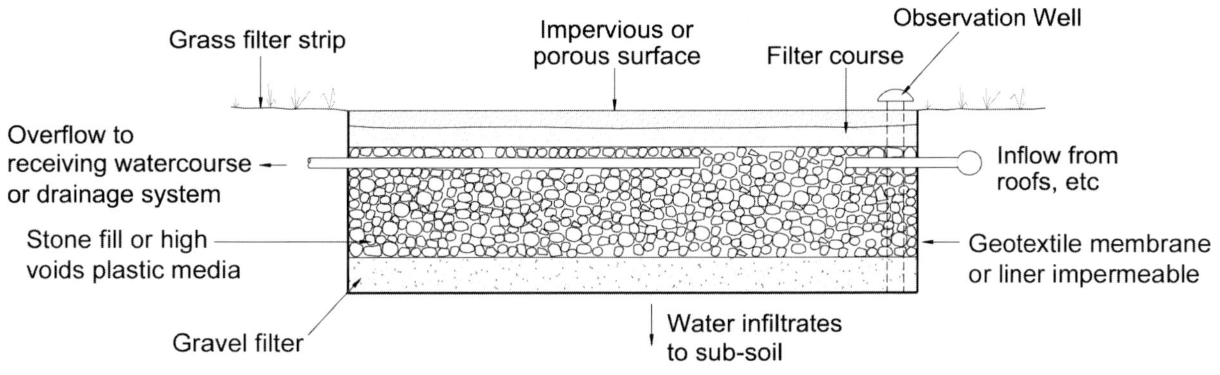

Figure G2 *Sub-pavement storage*

G2.2 Design principles

The volume of storage required needs to be designed for a particular runoff flow rate and volume, and consideration should be given to the possible implications of an extreme event occurring. The inflow and outflow perforated pipe systems should be designed to accommodate these flows, the proportion of voids assessed for the design depth (usually at least 300 mm) and an overflow system considered. Inflow rates should take short intense storm events into account, while long wet periods need to be used for volumetric analysis. It is likely that design return periods used are usually between 1 in one and 1 in two years, although individual circumstances may require either smaller or larger values to be used. This draws attention to the need to consider the likely performance and behaviour for a 30-year storm when using a rainfall runoff impact assessment method to assess total site drainage needs.

Where infiltration is being considered, current best practice is to use CIRIA Report 156 (Bettess, 1996), although many engineers still use BRE Digest 365 (BRE, 1991).

If the paved area is on a significant slope it is possible to introduce internal barriers along contours within the sub-base to retard the flow as it passes through the media. Considerable research has been carried out to test out the robustness of such systems and measure their benefits. In general, these systems have proved very effective, both hydraulically and in dealing with pollution.

CIRIA C521 (Martin *et al*, 2000a) provides limited technical guidance for the methodology for designing sub-pavement storage.

G2.3 Maintenance requirements

The design life of car parks is usually in excess of 10 years, so it is important not to risk regular flooding occurring as a result of the porous media or the distributing pipework becoming blinded or blocked with silts and sediments. The distribution pipework introducing the water into the media should be extensive, rather than just introduced at local points. Also, where it is perceived that there is a high risk of solids washoff, it is usually possible to route the first part of the collected flow through a silt trap or a bed or trench filled with coarse stone outside of the paved area. This acts as a good filter, and if it blinds up it is very easy to dig it out, wash it and reinstate it.

G3 Swales

G3.1 Description and purpose

A swale is a broad, shallow channel with vegetation covering its side slopes and base. Swales can be natural dry channels or man-made. Their shape is quite different from that of standard ditches in that they are wide-bottomed and have gently sloping sides. They are usually grassed (but sometimes can be designed to have a wet base) and therefore require regular attention to prevent them from becoming overgrown. Swales also provide some storage to attenuate the runoff and, in areas of sandy soils, some reduction in flows is also achieved via infiltration of the runoff into the subsoil. In addition, pollutants from the roads, especially solids, are partially adsorbed before the runoff reaches the watercourse. Figures G3 and G4 illustrate the shape and size of a typical swale.

Although normally associated with road drainage, the concept of swales can be used in industrial and commercial developments for roofs and car park runoff.

It is often difficult to introduce swales retrospectively owing to limited land availability. Their use in dense urban developments is rare for the same reason.

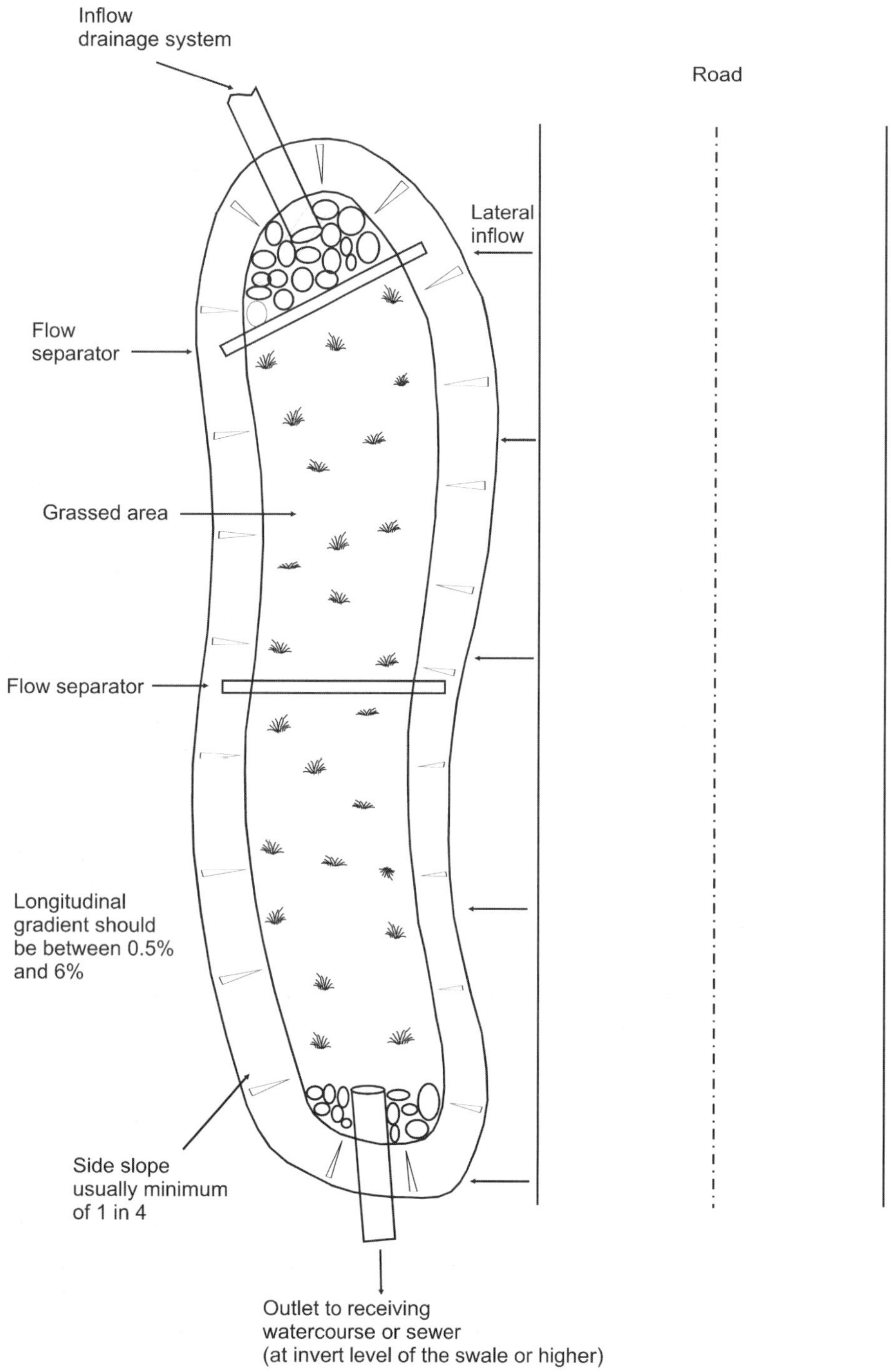

Figure G3 *Plan of a swale*

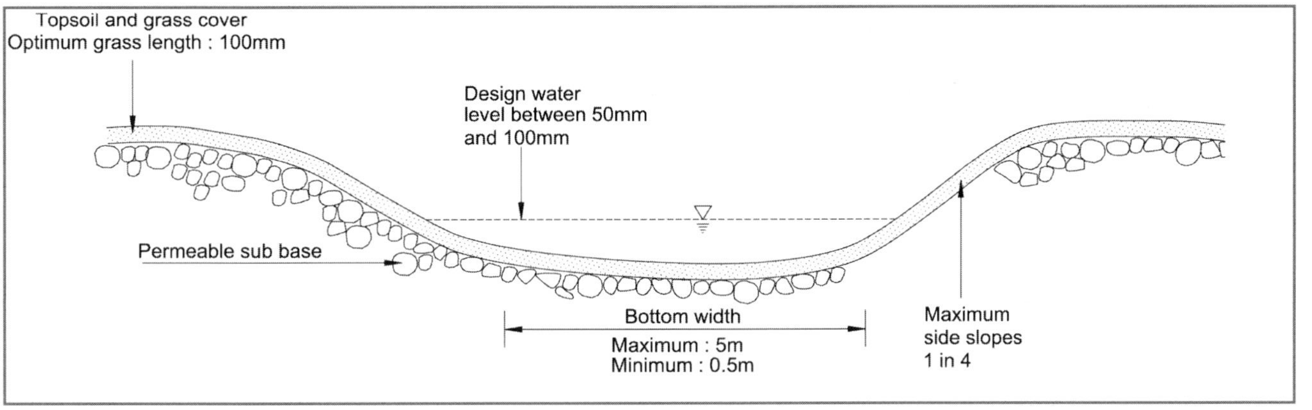

Figure G4 *Section through a swale*

G3.2 Design principles

To ensure that the grass remains healthy, a swale should be designed so that water does not stand for long periods. Longitudinal gradients must be low to prevent excessive velocities that can lead to erosion. Control weirs can be used to achieve the necessary gradient. Because depths of flow need to be low enough to prevent high flow velocities, the area served by a swale is generally small, with flows then being collected and channelled or piped. To prevent erosion at input of point sources, roads adjacent to swales should be designed without kerbs and gullies – a feature that can also reduce construction costs. The conveyance of flow through pipes from one swale to another also has the potential to cause blockage or erosion problems.

Although designed to operate with a low depth of flow, the outflow unit, if restrictive, can enable the full depth of the swale to act as attenuation storage for brief periods.

By installing a granular trench along the centre-line, a swale can also be used as an infiltration trench, if the soil type is suitable. If it is not, then a perforated pipe in the granular bed can be used. These additions significantly extend the usefulness of this method of drainage. Although design storms can be used to assess the hydraulics (both long and short storm events), the dry periods are equally important to evaluate in considering the design of the vegetation. This implies that a rainfall time series should be used. In summary, swales are very flexible units enabling infiltration, attenuation and treatment, but they do require a significant amount of room. They are effective in terms of amenity, both visually and from a general ecological perspective.

The concept of using check dams at regular intervals is recommended to ensure that a low flow channel does not develop.

Consideration should be given to using suitable grasses.

CIRIA C521 provides a worked example of designing a swale. This is in two parts, using both a one-year and a 20-year event. Both events use the concept of an average rainfall intensity to determine flow depths and velocities using a Manning's approach. The use of the 20-year criteria is to check on conveyance and velocity, while the one-year is to check on effectiveness of the swale to treat the runoff to improve the water quality of the runoff. The use of average rainfall intensity is suitable for determining depth, but volumetric assessment of storage requires the use of hydrographic techniques.

A criterion for treatment suggests that flows should have a velocity of less than 0.3 m/s. Other design parameters to ensure some degree of treatment is achieved are that depths of flow should be no more than 0.1 m. The maximum side slopes of swales should be 1:4. Where the swale is just being used as a hydraulic channel, slopes can be slightly steeper and velocities raised to 1.5 m/s. The longitudinal gradient should be between 0.5 and 6 per cent. The report suggests that where a swale is being used for impoundment, that it drains fully within two days. This is necessary to protect the healthy growth of grass. It is suggested that the grass should be kept fairly long (100 mm) to assist in trapping silts.

G3.3 Maintenance requirements

The maintenance of swales is primarily limited to grass cutting. Although very frequent (in comparison to most other drainage units) it is low cost and easily carried out. However swales may suffer from erosion or become water logged if poorly designed or used inappropriately.

Outflow structures and connecting pipework between swales (under roadway access) must be kept free from blockage.

G4 Tank sewer storage

G4.1 Description and purpose

One of the standard requirements by sewerage undertakers for accepting surface runoff from a new development into an existing drainage system is to specify a maximum discharge rate from the site. Where site runoff is being discharged to a river, the Environment Agency also usually requires discharge rates to be limited. At the beginning of a planning proposal a developer often has little idea as to the storage constraints, if any, that will be applied. This then has to be resolved, and often the design criterion used by the Environment Agency differs from area to area. This is due to differences in methodology and to specific catchment considerations. The discharge limit is often in the region of 5–7 l/s/ha (see Chapter 10 for details), although figures range from 1 l/s/ha to as much as 20 l/s/ha. *Sewers for adoption* (WRc, 2001) suggests protection against flooding of the development should be for 30 years, but the Environment Agency generally looks for a higher standard for the protection of the watercourse. Typically a 100-year standard would be applied.

To comply with the above criteria requires the provision of storage. Most small developments usually comply by putting in oversized sewers of 900 mm or 1200 mm diameter or occasionally box culverts of sufficient length to meet the storage volume needed. There are also some suppliers who specifically cater for this requirement by providing large prefabricated tanks or pipe lengths in PVC or GRP. Most storage is put in as on-line storage, though off-line tanks are occasionally used. Ponds are generally more desirable for water quality (partial treatment of the runoff), but historically they were rarely built due to the difficulty of getting ponds adopted. Tank sewers, subject to good construction practice, are usually acceptable for adoption by all sewerage undertakers. An additional factor, which influenced the choice of tank sewers, was the additional land that was available for development.

The alternative of an *in-situ* concrete structure usually only becomes necessary for very large developments; they tend to be constructed by sewerage undertakers on trunk sewers. The design principles for large structures are effectively the same as those needed for the smaller ones.

CIRIA C521 does not include tank sewers as a SUDS facility.

The principle features of a tank sewer are illustrated in Figure G5.

G4.2 Design principles

There are various issues to consider in designing storage for a site.

G4.3 Volume

The accepted approach is to use standard design storms within the Wallingford Procedure runoff model, as discussed in Chapter 10. Traditionally, storms of various durations have been run for the return period of interest and the largest storage volume predicted was selected. This is no longer considered best practice for providing storage for protecting the river against flooding, although it remains appropriate for determining site system performance. Although fixed discharge limits are specified, the designer should be aware that this is only achieved at maximum head. The head-discharge relationship should therefore be taken into account in establishing the actual volume needed at the stage of detailed design. The additional storage implied by this limitation can be allowed for in designing tanks using two-stage filling to minimise the storage volume.

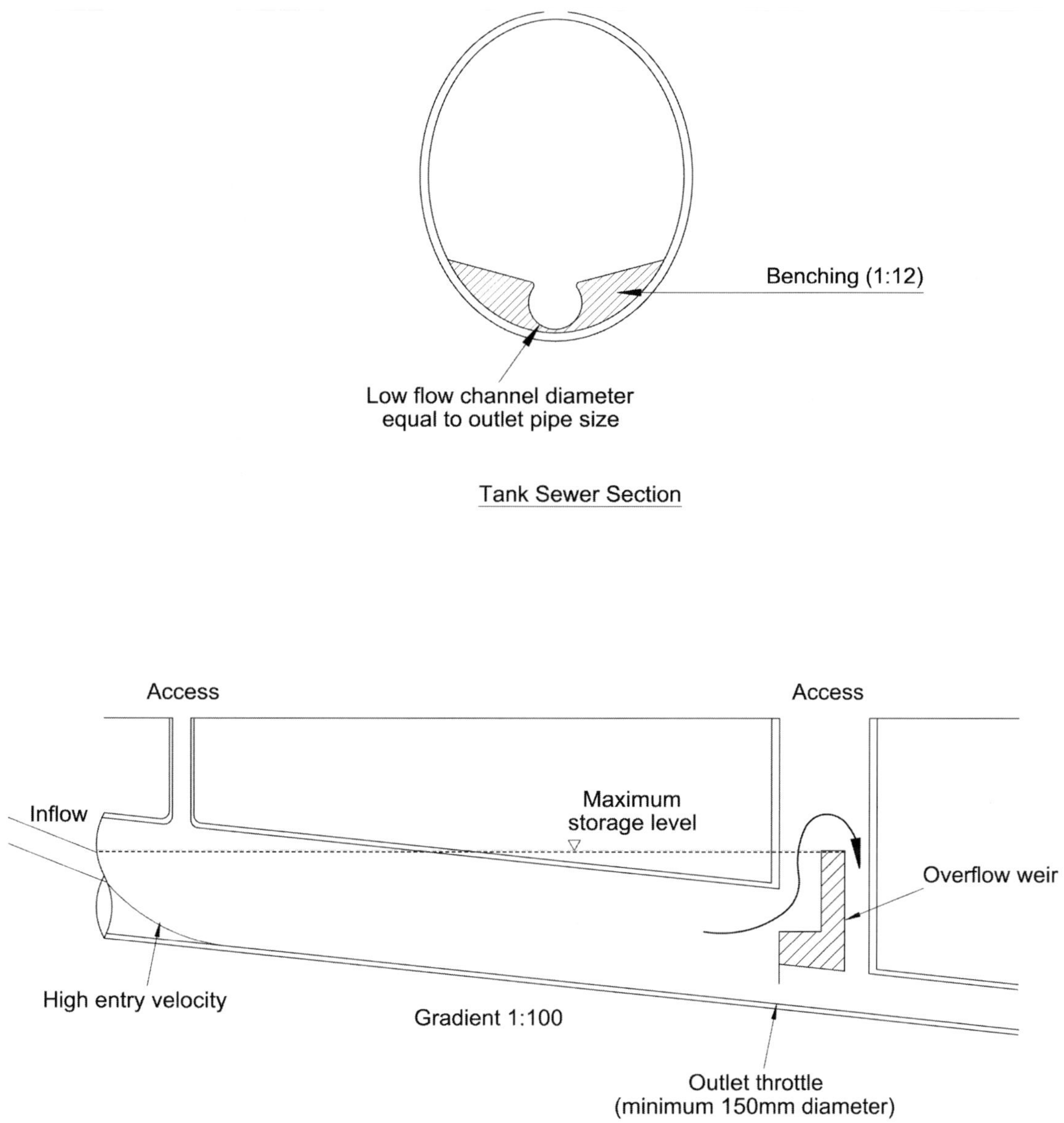

Figure G5 *Tank sewer*

G4.4 Topography

Topographical features should be taken into consideration to ensure that there is sufficient cover for the tank and that gradient is such that the tank will drain by gravity. Chapter 3 discusses the issue of topography.

G4.5 Geometry

Tank geometry is important in order to achieve self-cleansing. The entry into the tank should be designed so that a high entry velocity is created. To ensure that velocities are high for small events, it is recommended that a "dry weather flow" channel be created. The longitudinal gradient should be sufficiently steep to ensure velocities are in excess of 1.0 m/s.

G4.6 Operation

One of the principal problems of using tank sewers for developments is that the limit of discharge has been specified as low as 2 l/s. To achieve these low flow rates, theoretical orifice sizes (even when using vortex controlling units) are less than 150 mm diameter. This sewer diameter is officially the smallest that sewerage undertakers will adopt. Using a smaller orifice is often perfectly viable, but the smaller the unit, the greater is the risk of blockage. Another option, commonly pursued in other European countries, is to use a small pump, although this adds to the maintenance and running costs of the sewer system.

The use of vortex control devices is not favoured universally. Some units have been shown to be more accurate than others are in achieving the design flow rate. The limitation in accuracy is partly due to the importance of the correct alignment of the unit. Some sewerage undertakers would prefer a minimum orifice size of 200 mm diameter if a vortex or other control unit were used.

G4.7 Design horizon

There is some concern that the life of some tank sewers, particularly galvanised iron structures that have low construction costs, may be less than that of traditional concrete structures. To date there has been insufficient feedback to be more specific about the limitations of the options available.

Another issue is the potential build-up of sedimentation. The number of tank sewer units around the country is growing rapidly as time passes and, unless maintenance is frequent, it is possible that the effective storage volume may be significantly reduced over time.

G4.8 Access/safety

Care must be taken in designing access and ventilation for man entry in accordance with normal CDM practices (see Section 9.2.9). In addition, consideration must be given to situations such as blockage and the access and activities necessary for emptying a tank in this state. Building in a penstock to draw down water out of the tank is a standard solution for larger tanks.

G4.9 Extreme events

Although tanks are designed for a specific return period, more extreme events may need to be catered for. This involves the prediction of flooding, its location and whether there is a need for, and the size of, an overflow structure.

G4.10 Maintenance requirements

A range of mechanical and hydraulic methods is used in the maintenance of large tanks. However, it is thought that tank sewers for developments are very unlikely to be of a size that requires these features to be built in. In theory, all tanks should be visited once or twice a year and deposited sediment flushed out with high-pressure jets or manually excavated. In practice, tank maintenance can be quite infrequent and there is believed to be no detailed information on tank cleaning requirements for small tank sewers at present.

G5 Roof storage

G5.1 Description and purpose

Roof storage can either be provided using flat roofs that automatically attenuate runoff significantly, or by using oversized gutters. The latter have gained credibility in Australia where discharge is "diffuse" along specifically chosen gutter lengths or the water is stored and used for toilet flushing. Roof storage is rarely considered in UK where houses are normally designed with pitched roofs. In tropical countries, where rainfall intensities are severe, flat roofs can be used to reduce the rate of flow into receiving sewers, although there is the obvious disadvantage of increasing the risk of significant roof leakage and water damage within the property.

A related, but rarely used, option in the UK is the use of "green" roofs. Although it has achieved a degree of acceptance on the "green fringe", the concept has some way to go before it receives general approval and enters wider use. The advantages are chiefly the improved thermal efficiencies of the dwelling, the reduced and attenuated rainfall runoff, and the partial treatment of the rainwater (particularly appropriate in some industrial areas).

G5.2 Design principles

The design principle is a simple one of reservoir design and sizing the outflow throttles using hydrograph techniques. The implications of a blockage are serious, so careful attention should be given to providing safety features that prevent storage of excessive depths of water. Multiple roof outlets should be designed to limit the risk of blockage.

CIRIA C521 does not consider the use of roof drainage as a SUDS technique.

G5.3 Maintenance requirements

Blockage can arise from the growth of moss, particularly on stone chippings. Leaf detritus is an problem in many suburban locations. The surface is usually sealed with a bitumastic material, which has a relatively limited design life compared to roof tiling.

G6 Infiltration basin

G6.1 Description and purpose

Infiltration basins often consist of one or more basins in a variety of shapes that store water and allow it to infiltrate into the subsoil. Locations that have heavy soils provide little scope for infiltration, although arrangements can be made to maximise the opportunities available. A typical infiltration basin is shown in Figure G6. They are used to supplement infiltration to groundwater and to reduce or prevent flows passing downstream. They are therefore used only where groundwater is not threatened by pollutants.

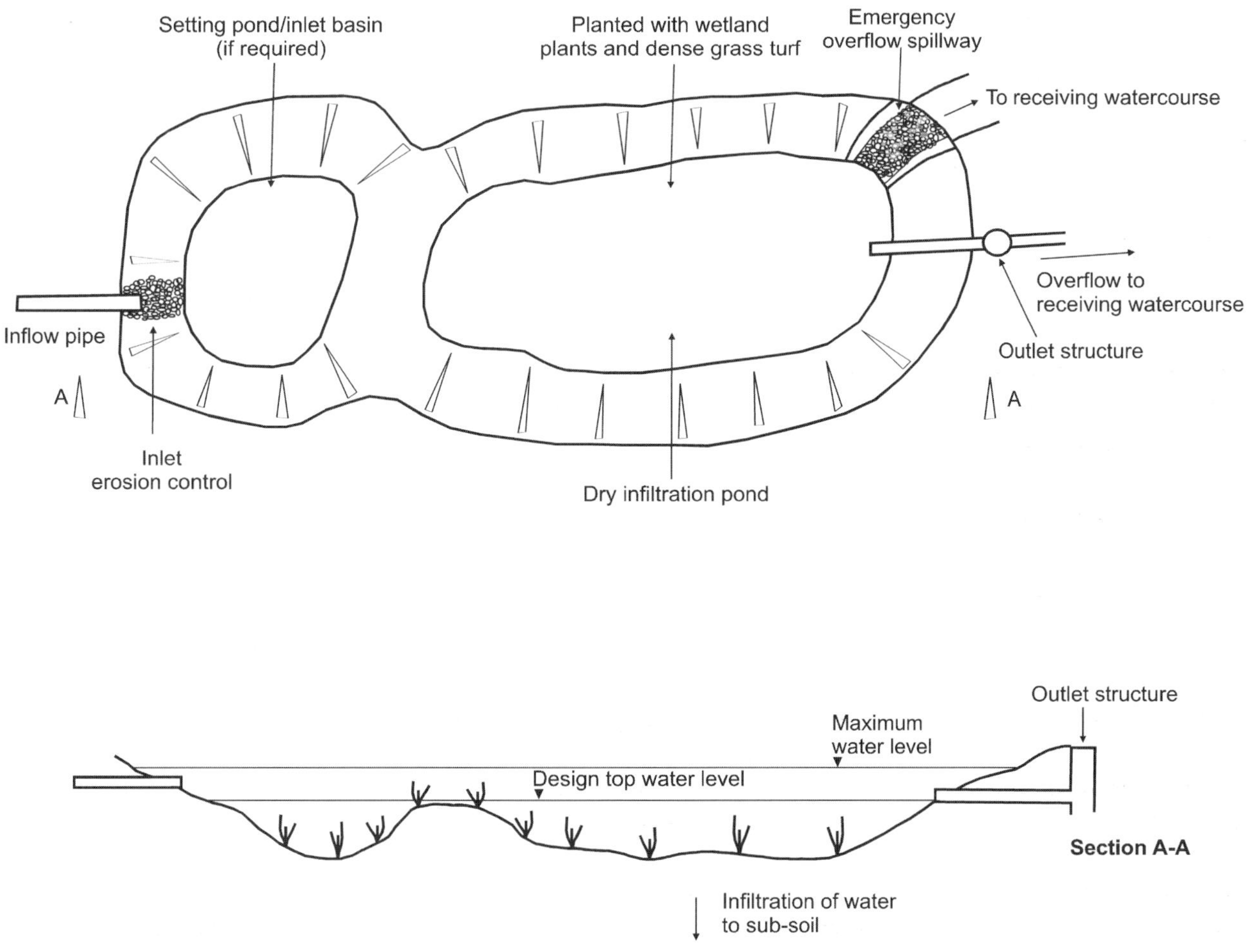

Figure G6 *Infiltration basin*

G6.2 Design principles

No design standards exist at present for infiltration basins. Among the points that need to be considered are:

- infiltration rate
- provision of an emergency spillway
- ground slopes.

The rate of infiltration, together with the surface area available for inundation, dictates how deep and how long surface water will pond before dissipating. The infiltration rate is difficult to determine, as fine sediments will be deposited and accumulate over time. In addition, high groundwater levels in winter may affect percolation, although the basins should not be built to depths that would meet the groundwater. The design should be based on maximum inundation times (say once a year) related either to convenience (if there is an alternative use of the basin area) or the physical effects on grass if vegetation is an issue. A time series approach is therefore most appropriate for assessing depths and periods of inundation.

Provision for an emergency spillway and flood route should be allowed for in the design of the infiltration basin. The emergency spillway will allow water to pass out of the infiltration basin when the storage capacity of the basin has been exceeded. Return periods would normally consider a minimum of a 100-year event for the design of these elements.

CIRIA C521 provides advice as to where and how the basin should be constructed. The advice on design is limited to stating that half the storage volume should drain within 24 hours and that this should be applied using a two-year or shorter return period storm.

G7 Maintenance requirements

Maintenance may involve grass cutting or scarifying, depending on the method of design of the structure.

G8 Paved surface flooding

G8.1 Description and purpose

Flooding of paved areas is used as a convenient method for temporary storage of excess surface water, usually on an infrequent basis. This allows the option of providing smaller storage facilities in the ground and having water backing up to the surface during extreme events. Car parks are usually selected as suitable locations. To make use of car parks or roads, it is important to carry out a detailed assessment of ground levels and the backing-up mechanism used to flood the area. It is suggested that flooding should not take place more often than once a year and the maximum depth of storage should be 150 mm for a 1 in 10-year or 1 in 30-year event. These criteria should be modified according to the level of service needed for each situation. For example, the flooding of parking of emergency vehicles may not be tolerated at less than a 1 in 30-year frequency. A level of 150 mm is unlikely to cause damage to any vehicles. It may be preferable or necessary to consider storing additional volumes in adjacent landscaped areas.

With the trend towards requiring consideration of the impact of more extreme events (100 years), car park and road flooding is an obvious and relatively easy option with which to manage extreme events. It requires the employment of fairly accurate modelling techniques and careful use of road alignment and landscape planning.

G8.2 Design principles

Detailed modelling of the sewer network and the ground levels using hydrograph methods ensures a safe assessment of flood depth prediction. Use of a rainfall time series allows an accurate frequency assessment to be made of the depth and duration of flooding. CIRIA C521 does not consider the use of car parks and road flooding as a SUDS option.

G8.3 Maintenance requirements

There are few maintenance implications except that some fine sediment may be deposited on the surface. Regular waterlogging of the pavement sub-base is regarded as poor practice for normal highway construction. It is therefore important to consider this issue. However, in the light of the acceptance of sub-pavement storage of water, it should only affect design of the pavement rather than prevent the use of temporary surface flooding.

The maintenance implications are minor by comparison to many other SUDS options and the main issue is the ability to design for an adequate level of service and manage the intended flooding.

G9 Detention basin

G9.1 Description and purpose

Dry detention basins are used to store water for brief periods. Detention basins are differentiated from retention ponds by the fact that they are normally dry and operate only during wet periods to constrain the rate of flow downstream. They are often very large, particularly overseas, where large volumes of water are stored to reduce flow rates from areas with fast response times draining large urban or semi-urban catchments. Dry detention basins are sometimes also used as leisure areas such as football pitches or parks. They are also used in commercial or industrial areas, where a small basin can be used to provide surface flow balancing.

The design of the site, particularly with regard to its use during dry weather, is very much a function of the frequency with which inundation takes place. Areas can be designed with two levels to allow frequent and infrequent flooding. Careful design is needed where grassed areas are planned, as the frequency and duration of flooding must be limited to ensure the survival of the grass.

One of the unfortunate aspects of such basins is the tendency for debris and litter to accumulate. Trash screens are needed at outflow structures to prevention blockage at this point or further downstream in the pipe system. They also prevent children from entering the outfall structure and pipework.

Detention basins are found in the lower end of the stormwater management train of the SUDS philosophy.

G9.2 Design principles

Design of pond storage and the throttle size can be easily carried out using deterministic models with design storms or a rainfall time series. Consideration should be given to extreme events, particularly for larger ponds. Detention basins should incorporate a provision for an overflow and the flood route. It is suggested that design of this structure should be to a 100-year return period. Outflow rates are a function of downstream limitations. Alternatively, outflows might aim to replicate the equivalent greenfield runoff.

Consideration should be given to both the design of a dry weather flow channel as well as the design extreme event. If the structure is large, it might also be appropriate to design landscaping related to frequent events of one- or two-year return periods.

Where flood volumes exceed 25 000 m^3, the 1975 Reservoirs Act comes into effect. In these circumstances, return periods of 300 years up to the "probable maximum flood" need to be designed for. This means that great care is needed when considering the effects of overtopping and in the provision of spillways.

Normally all orifices are designed with screens to prevent debris blocking them and to comply with health and safety aspects. However, screens allow litter to be deposited against the bars, which reduces the outflow rate. To minimise this effect, design of the flow entry structure should provide a large surface area of screen, possibly at various levels, with cleaning platforms between levels. Other safety aspects, such as depth of water and velocities, should be considered.

CIRIA C522 (Martin *et al*, 2000b) considers the use of detention basins for both water quantity and water quality use. For water quality benefits, it advises that the outflow time should extend to 24 hours to maximise sedimentation and partial treatment process. Hydraulic design criteria are limited to the suggestion that the pond should be sized to a function of the volume V_t. The book suggests that an initial settlement basin – representing 10–25 per cent of the plan area of the whole structure – be built to trap coarse sediment and debris. This will often be constructed as a wetland area and will require maintenance or partial rebuilding about every 10–25 years. C522 also gives some key points on construction issues. The book points out that detention basins are only part of the SUDS philosophy and that source control options should have been considered before using this type of structure. C522 highlights the significant ecological and other amenity opportunities that exist for this type of structure.

Another reference for the design of inlet, outlet and overflow structures for larger basins and ponds is CIRIA Book 14 *Design of flood storage reservoirs* (Hall *et al*, 1993).

G9.3 Maintenance requirements

The outflow screen needs to be kept clean. Screens with shallow angles can be self-cleaning to a certain extent. If part of the detention pond is embanked, then care is needed to check for slippage, rodent damage, and settlement. Large sites serving streams are likely to provide a certain amount of debris, which will need to be removed after events.

Regular inspection and management of the vegetation should take place.

G10 Retention ponds

G10.1 Description and purpose

Retention ponds can be differentiated from detention basins in that they are permanently wet. Retention ponds often have a longer storage retention time, as greater emphasis is often placed on the potential for water treatment. A typical illustration of a retention pond is shown in Figure G7.

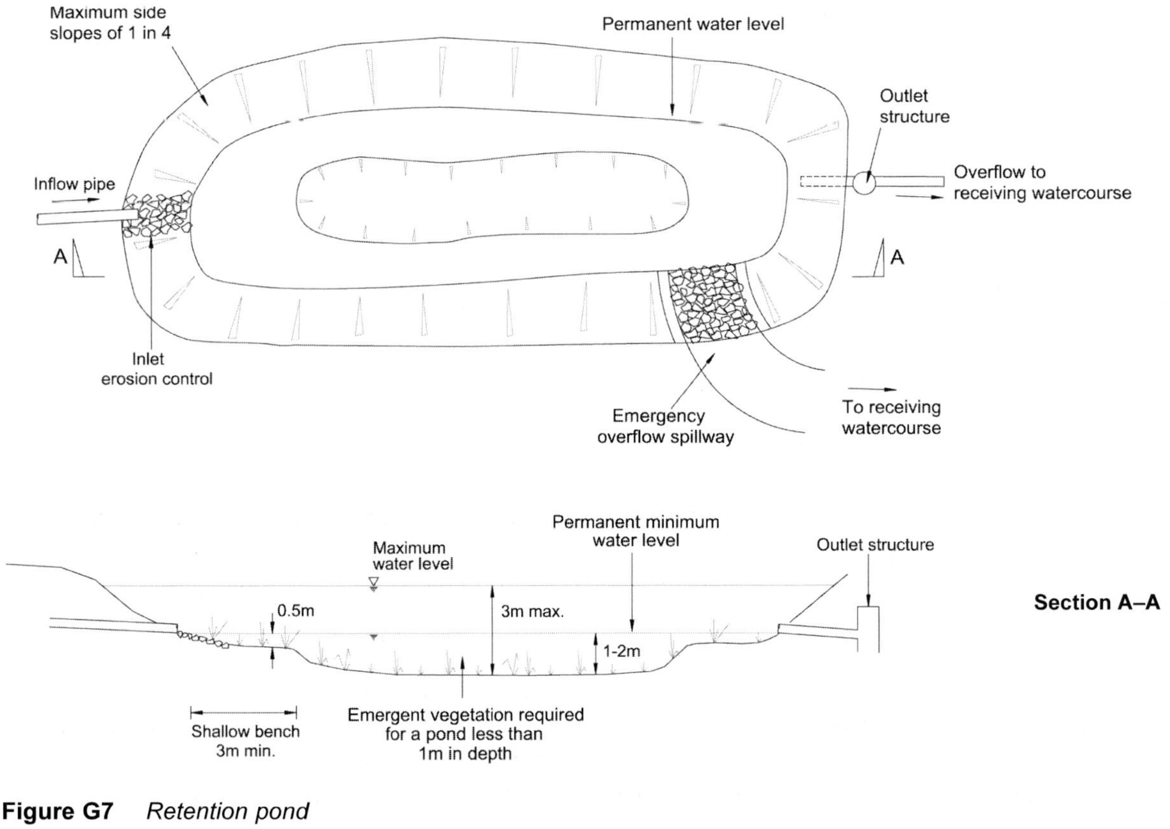

Figure G7 *Retention pond*

Retention ponds have the advantage of generally being aesthetically pleasing and enhancing the environment as well as providing water quality benefits. The principal disadvantage is the higher maintenance requirement of managing the pond and regularly harvesting the plants. Safety is also a significant issue, and care must be taken to provide a shallow shelf at the edge of the pond and plants that grow to form a barrier to prevent easy access into the water. Easy access must, however, be provided for getting out and away from the pond side.

G10.2 Design principles

Design of the pond should be for both hydraulic and water quality aspects.

Hydraulic design is similar to that for other basins in that the routeing effects are a function of surface area, storage and outlet control. Hydrograph techniques should be used for design. The design return period for the structure will depend on its use. Hydraulic aspects would consider 30- or 100-year events, while water quality design will consider the one-year event or a rainfall time series. Consideration must be given to performance under extreme events and flood flow routes. In addition, the frequent event characteristics could benefit from specific design. Greenfield runoff is normally minimal for small events in summer, whereas urban runoff provides concentrated polluted runoff in these circumstances. The use of retention ponds allows a designed infiltration freeboard to be incorporated, to try to replicate the greenfield condition and so protect the receiving stream. A rainfall time series analysis allows evaluation of the required freeboard and infiltration rates and gives an indication of the level of low-flow river protection provided.

The risk of stagnant ponds is significant in urban areas where nutrient sources are plentiful. Aeration, using either particular species of aquatic plants or fountains, reduces the risk of eutrophication. The design of ponds in terms of aquatic planting requirements is a specialist area and is not covered in this publication.

Careful alignment of the inlet and outlet structures and the use of islands should prevent short-circuiting of flows through the pond. Table G2 gives typical pollutant removal efficiencies that are achievable for various basin sizes.

Table G2 *Pollutant removal efficiency for size of pond structure*

Storage basin size	Pollutant removal efficiency
US – Schueller, 1987	
0.5 inches of runoff/acre	60–90 per cent solids removal 35–90 per cent phosphate removal
2.5 times runoff from mean storm	75 per cent solids removal 55 per cent removal
UK – Hall et al, *1993*	
1 per cent of catchment (area) 150 m^3–250 m^3/impervious ha	80–90 per cent solids removal 50–60 per cent removal of soluble pollutants
Vol storage / QBARu = 4 to 6 Detention vol > 100 m^3/effective ha (where QBARu is mean annual flood)	Maximum removal of all pollutants

CIRIA C522 advises that the design storage volume should be equal to $4 \times V_t$ to achieve effective treatment of the runoff. It suggests that retention ponds should not necessarily accept all flood runoff and that larger events would also be served by other structures such as detention basins. The aim would be to maximise the advantages of treatment without requiring massive pond sizes to cope with all storms.

Advice is provided on the construction shape and depth and also the aquatic planting requirements. Aquatic plant species should range from emergent to floating and submerged. Emphasis is placed on planting with plants that are appropriate for the local climate, soil and water. Information on inlet, outlet and flow control structures can be referred to in CIRIA Book 14 (Hall *et al*, 1993). Maximum opportunity should be taken to enhance both ecological and aesthetic benefits that these structures afford.

G10.3 Maintenance requirements

Maintenance of retention ponds involves both aquatic plant management and the physical checks to ensure the structures remain sound. Some harvesting of reeds or replanting of reedbeds may be necessary about every 10 years. The same checklist of maintenance activities should be carried out as those given in Section G9 on detention basins, although with the emphasis primarily on vegetation management. Recent experience in the UK with a variety of foreign plant species that, once introduced to ponds, necessitate extremely costly management, underlines the importance of protecting the ponds and using only appropriate vegetation.

Sediment removal might require more frequent attention to prevent the accumulation of high concentrations of micro-pollutants. Research has not yet established whether this is a significant issue, nor the frequency with which sediment should be removed.

G11 Permeable pavements

G11.1 Description and purpose

Permeable pavements allow precipitation to infiltrate through them, thereby reducing runoff rates. The most common form of construction is the use of highly porous blockwork, though concrete cellular or plastic media filled with aggregate/shingle are often used too.

The term "porous pavement" refers to the use of a no-fines top dressing layer on roads. The material allows water to drain to the side of the road without causing traffic spray. It also reduces tyre roar and so has considerable benefits for fast roads passing through urban areas. This surface is really a different category to the other forms of pavements that pass water through to the sub-base material.

Permeable pavements are generally limited to car parking areas due to the reduced loading capabilities of waterlogged sub-base materials that are specifically designed to have a high voids ratio.

Permeable pavement substrate removes much of the pollutants by adsorption and microbial action from the runoff and allows recharge of local aquifers if a liner is not used. Liners are often specified to prevent the possible contamination of groundwater.

Porous pavements can become clogged and there is general uncertainty in the industry as to the risks in this area. This is probably the main constraint in their general use. Although designed with very high percolation rates (> 4500 mm/h), blockwork can become virtually impermeable once oil and dust have been on it for some months. There is now a move away from pervious blocks and using designs that specifically allow percolation to take place between the blocks.

Figure G2 shows the majority of the features associated with a permeable pavement construction.

G11.2 Design principles

There is little information on the hydraulic characteristics (inflow, outflow and losses) from pavements with sub-base storage. Information is also limited on practical issues of translation of flows, density of pick-up pipework and other details that engineers need to design the surface hydraulics of a pavement.

Permeable pavements have often been found to have short life-spans. The poor performance of permeable pavements and other infiltration methods is often attributed to defects in the design, poor construction techniques, low-permeability subsoils and lack of adequate preventative maintenance. Important design criteria that should be taken into account are detailed below.

1. **Site evaluation**. The site evaluation should include the following:

- soil borings should be taken at least 1.5 m below the stone backfill to assess:
  - soil permeability and porosity
  - depth of the water table during the winter
  - depth to the bedrock/impermeable surface
- the site should not have a slope greater than 5 per cent
- a minimum soil infiltration rate of 15 mm/h is suggested where infiltration is intended.

2. **Traffic loading and conditions.** The permeable pavement should be carefully designed to suit the imposed loading conditions. They can be designed for relatively heavy loads, but are not suited to high-impact use such as main roads. The use of sand, salt and deicing chemicals in the winter should be restricted.

3. **Construction**. The construction of a permeable pavement will typically comprise:

- porous asphalt course between 50 mm and 100 mm thick or porous concrete blocks
- filter aggregates
- a reservoir course comprising stones between 40 mm and 80 mm in diameter
- a filter fabric.

The permeable pavement should be constructed using light equipment to prevent the compaction of the soil. In some cases it is recommended that a vegetative filter strip 5 m wide be planted around the perimeter of the porous pavement. This filter strip protects the surface sediments in the runoff that might drain towards it.

Where possible, permeable pavements should be constructed only after all major earth-moving and landscaping on site have been completed, to minimise contamination and sealing of the surface. The surface can be protected by covering with geotextiles if necessary.

G11.3 Design method in CIRIA C522 and C582

The two books (Martin *et al*, 2000b and Pratt *et al*, 2002) refer to a number of standards for the design of pavements. They detail all the key issues that need to be considered for design and summarise the surfacing methods used and the construction materials needed.

There is no information yet on details of pipe sizing, media size and grade, and the voids ratio for calculating the design hydraulic performance of the pavement.

If the sub-base under permeable pavements is lined, then outflow pipework is needed at the bottom of the sub-base to drain it. If it is not lined, then infiltration is encouraged, and a drainpipe is placed at a high level within the sub-base to protect the surface from waterlogging and frost-heave effects. The books point out that research has shown that temperatures rarely drop below freezing in the sub-base in the UK.

The books emphasise the need to consider loading limits in designing the pavement.

G11.4 Maintenance requirements

Maintenance of permeable pavements should include vacuum sweeping from two to four times a year. High-pressure hosing to prevent the pores in the top layer from clogging should follow the vacuum sweeping. After completion, the pavement should be inspected several times in the first few months, followed by regular annual maintenance.

G12 Soakaways

G12.1 Description and purpose

Infiltration in the form of soakaways is a well-known form of stormwater disposal. With the advent of a greater consciousness of the importance of sustainable lifestyles, methods of using infiltration have been addressed in some depth. The distinction between soakaways and other forms of SUDS is that more is known of their performance and the technical guidance is established and accepted. This is primarily because design criteria has been established and proven over many years.

Soakaways are used primarily to drain roofs in areas where groundwater levels are low (throughout the year) and the ground is permeable. They have been used for road drainage as well. Over time (usually within 10–20 years) they tend to blind up. Very careful design allied with regular high levels of maintenance has shown that soakaways can be used for roads.

Current best practice on infiltration measurement and design of soakaways is defined in CIRIA Report 156 (Bettess, 1996). However, BRE Digest 365 (BRE, 1991), which also deals with this subject, is still used by many in the industry and is also used in the illustration in the CIRIA SUDS guides.

The importance of soakaways cannot be over-emphasised, as infiltrating all the runoff from roofs deals with up to half of the impermeable surfaces in urban areas. They can therefore make a major contribution towards reducing the impact of rainfall runoff and increasing groundwater recharge.

Soakaways are constructed below ground and come in several forms. They may be constructed from excavations filled with stone or from pre-cast concrete rings (with holes) with a backfill around it of aggregate. A typical example is shown in Figure G8.

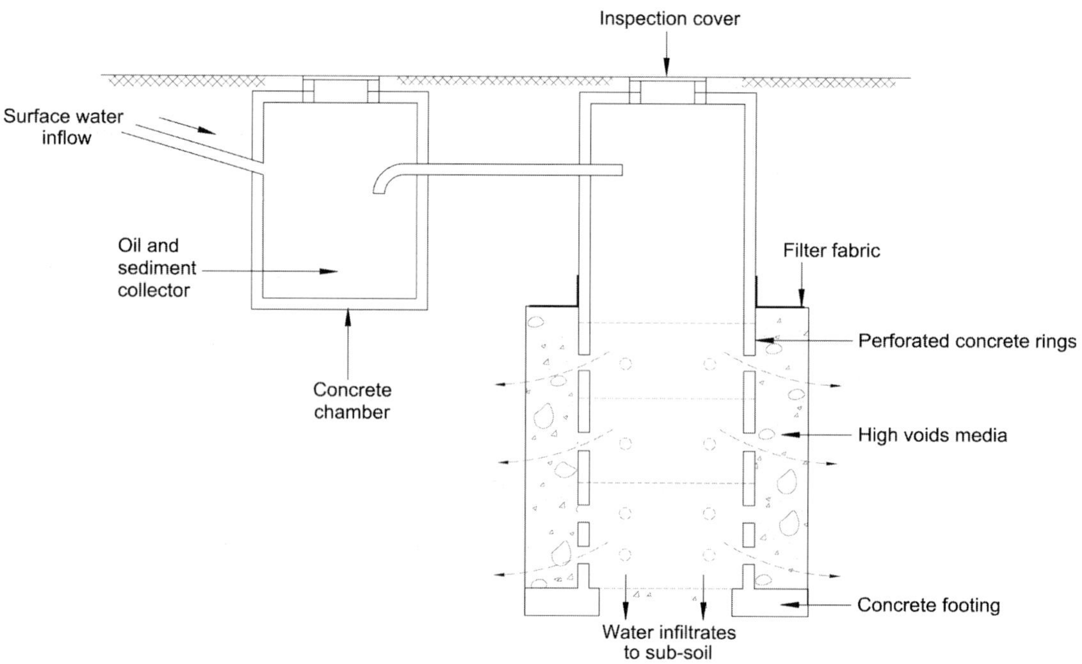

Figure G8 *Soakaway*

Runoff from the roof or roofs is stored in the soakaway and allowed to infiltrate into the soil. In some cases, a chain of linked soakaways is constructed using interconnecting pipes. To protect against local flooding, a high-level overflow pipe is sometimes designed and led to a discharge point or another part of the drainage system.

It is recommended that the inflows to soakaways are fitted with sediment traps and, where necessary, oil interceptors, where surfaces other than roofs are being drained.

Soakaways should generally be used only where the soil has a low silt and clay content.

G12.2 Design principles

The design principles for soakaways are defined in the CIRIA and BRE references mentioned above. In outline, it is necessary to carry out infiltration tests at a sufficient scale and over sufficient time that an accurate estimation of the infiltration rate can be established. In addition, the range of the groundwater level, especially in winter, needs to be assessed. The Environment Agency recommends that soakaways should not intercept the groundwater table.

The head of water is an important part of the infiltration capacity of the structure, with the infiltration rate decaying in an exponential manner. The use of linked soakaways allows a larger surface area for infiltration for a given volume of excavation and reduces the weight of the construction units. There is however a trade-off between this and the volume of storage needed.

Hydraulic design of soakaways is best achieved using a rainfall time series approach, though design events can be used. The design return period should be in the region of 10 years, though the location and the dwelling being served may allow the use of a different criterion.

Although the method of approach is now well established, there are many instances when the evaluation of the infiltration rate has been seriously over-estimated. A conservative approach is generally advised in the assessment of the size required for a soakaway.

G12.3 Design method in CIRIA C522

CIRIA C522 treats soakaways and infiltration trenches together, as the principles involved are the same. It points out that trenches are easier to construct and have the advantage of higher surface-to-volume ratios, increasing the percolation capacity.

Their potential impact on groundwater should be considered. The water table should be 1–2 m below the bottom of the unit and they should normally be sited at least 5 m away from any building or road. C522 also suggests that (trench) depths are kept to a minimum to maximise the flow path to the groundwater table. However, as the head of water is an important element (CIRIA Report 156), this also reduces the efficiency of the unit.

The book does not detail a design methodology, as reference is made to the two main national guides on the subject. It does, however, advise that design of soakaways is usually based on a 10-year event.

Advice is provided on construction details, including the use of geotextiles to protect soakaways from blinding. It suggests that a 225 mm inspection pipe is included in a stone-filled soakaway to enable its performance to be monitored.

G12.4 Maintenance requirements

The CIRIA guide advises on annual inspections of soakaways to determine performance and to carry out maintenance. The long-term maintenance of soakaways can be difficult unless provision for access and cleaning is made as part of the design. If the performance of the soakaway deteriorates, then often the whole structure has to be replaced, as the soil around the structure can blind up. The use of sand as a filter has been incorporated into some designs. This allows the sand to be removed and replaced, which significantly extends the life of the unit. Any sediment trap or oil interceptor used should be regularly cleaned. Sometimes, high-pressure hosing can be used to remove sediment that may be clogging the filter material, but this type of jetting equipment should be used only after careful consideration of the risk of driving silts and fine material into the soil or the surrounding granular material.

Soakaways attract the ingress of tree roots. It is not clear whether this is a problem. Trees take up large quantities of water in warm conditions and the roots tend to provide flow paths for water into the surrounding soil. However, structural damage and deformation also can result.

G13 Infiltration trench

G13.1 Description and purpose

Infiltration trenches are often used as an alternative to soakaways. The principal problem in their use is their location in high-density developments, as it is generally advised that they should not be within 5 m of any structure or road. They have the advantage of ease of construction compared to soakaways. Their use and limitations are effectively the same as for soakaways discussed in the previous section.

G13.2 Design principles

Typically, infiltration trenches are between 1 m and 2 m deep, although they can be as much as 4 m. They are backfilled with stone aggregate or plastic media (which has the advantage of having a high voids ratio) and lined with a geomembrane. A geotechnical investigation should be carried out to assess the feasibility of using an infiltration trench. Infiltration trenches should generally only be used where the soil has a low silt and clay content. Nevertheless, they can provide significant attenuation and balancing effects as part of a total system network even in clay soil conditions. They have the additional benefit of preventing clay shrinkage and adding moisture to the ground in dry conditions.

Flows can be introduced as sheet flow along their length though it is more normal to have piped input points. The former is generally associated with the traditional roadside French or filter drain, which is covered in Section G14.

To prevent potential contamination of the groundwater, the bottom of the trench should be at least 1 m above the level of the water table. Runoff with high sediment or hydrocarbon loads should be diverted away from the infiltration trench to prevent it from becoming clogged. A vegetated filter strip at least 5 m wide should be established adjacent to the trench to capture coarse sediments. The filter strip should be graded at between 0.5 per cent and 15 per cent so that the runoff enters the trench as sheet flow.

To minimise compaction of soils during the construction of the trench only light equipment should be used. A geomembrane, which acts as a filter, should be placed around the sides and bottom of the trench and also 0.3 m below the trench surface. This filter prevents the ingress of soil from the sides of the trench, which can clog the aggregate. A typical infiltration trench is shown in Figure G9.

A site-specific trench depth can be calculated based on the soil infiltration rate, aggregate voids ratio, and the trench length and width. Typically, stone aggregate between 25 mm and 75 mm in diameter is used to backfill the trench. This provides a voids space of up to 40 per cent. An observation well should be included in the design to allow the monitoring of water levels. This usually takes the form of a 150 mm- or 225 mm-diameter PVC pipe attached to a footplate at the bottom of the trench. Consistently high water levels in the observation well may indicate the need for maintenance.

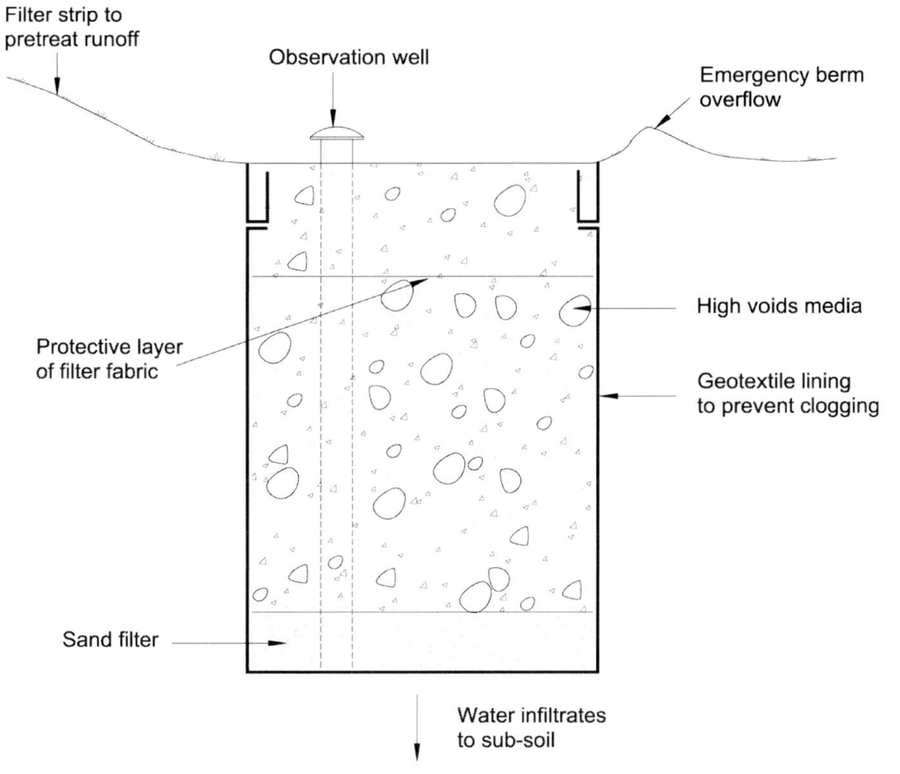

Figure G9 *Infiltration trench*

G13.3 Design method in CIRIA C522

The guide deals with soakaways and infiltration trenches together. Reference should therefore be made to Section G12.3.

G13.4 Maintenance requirements

The main objective in the maintenance of an infiltration trench is to prevent the clogging of the trench by sediment, rendering it ineffective. After large storm events the trench should be inspected and any debris removed. Thorough inspections of the observation wells should be carried out at least annually. These inspections should include monitoring of the time it takes for the trench to drain. Where filter strips are an important element, these should be inspected for damage or erosion after large storm events.

G14 Filter drains

G14.1 Description and purpose

Filter drains, commonly referred to as French drains, are usually stone-filled trenches adjacent to roads. They have a porous pipe in the bottom of the trench to pick up the road runoff, convey and discharge it to suitable low points. They have been used for many years, but their benefits are only recently being fully appreciated.

They have numerous advantages. There is no need for kerbs and gullies to be built, runoff is attenuated and some infiltration is also achieved. The stone media acts as a treatment zone, reducing pollutants in the outflow, particularly concentrations of heavy metals, by around 80 per cent.

Recent years have seen the introduction of variants employing geotextiles and other prefabricated units, which reduce the cost of construction. To the low-level perforated pipe has been added a high-level pipe, to protect the sub-base. There is also the possibility of removing the low-level pipe (to encourage infiltration), although the latter is not accepted practice due to the potential effects on the road sub-base.

G14.2 Design principles

Trenches are usually designed volumetrically to store 10 mm of rainfall. The low-level drain is designed with a capacity to drain the trench within 24 hours. Although this tried and tested design criteria is effective, a more complex approach of specifically modelling the interaction of the volume of storage, infiltration rate, and flow capacities of pick-up pipes would allow optimisation of the design and analysis of water levels in the trench.

Consideration should be given to the risk of blockage (failure) and the prevention of standing water on the carriageway.

G14.3 Design method in CIRIA C522

The criteria for hydraulic design are given in the guide. C522 provides details and discussion on construction issues and has information on the hydraulic and water quality benefits of using these units.

G14.4 Maintenance requirements

Filter drains should be maintained and inspected in a similar fashion to infiltration trenches. One additional aspect is the risk of scattering of stones across roads and also the problem of vegetation growing over the surface of the trench. This implies that regular management of the units is appropriate.

G15 Filter strips

G15.1 Description and purpose

Filter strips are areas of vegetation, usually a swathe of grass at least 5 m wide, that are often used to border another drainage facility such as a pond or a swale. The main objective of filter strips is to slow down runoff and to provide it with some degree of pre-treatment.

Their use in Britain has been limited due to the implied land take necessary to use these units.

G15.2 Design principles

A filter strip's effectiveness is dependent on the following:

- the amount of sediment reaching the filter strip from the surrounding area
- the period of time that the runoff is held by the filter strip. This is a function of the width of the strip, and the type and the condition of the cover used. In general, uniformly shaped filter strips comprising dense, healthy and deep-rooted plant species are the most effective at attenuating runoff and trapping sediments
- the soil's infiltration rate. Soils with high infiltration rates trap and hold more dissolved pollutants than soils with lower infiltration rates
- surface and slope uniformity. Small depressions and rills in the filter strip may concentrate the flow and reduce the strip's effectiveness. If the surface of the filter strip is not uniform, small rivulets and channels can form in the strip. These can decrease filtration effectiveness and carry sediments and pollutants through the filter strip too quickly for them to be removed effectively
- gradients should be mild to prevent risk of erosion
- plant species used in constructing the strip
- maintenance of the filter strip.

In general, a wider, uniformly shaped strip is more effective at stopping or reducing pollutants than a narrow strip. The effectiveness of the filter strip is also dependent on plant selection. Plants with fibrous root systems that form a dense mat of vegetative material (as opposed to those that grow in clumps) are generally to be preferred. Consideration should be given to the soil fertility and pH, soil drainage, time of seeding and species types to aid establishment. Before seeding, the land should be landscaped to allow flow to run over the strip in a steady, uniform manner.

G15.3 Design method in CIRIA C522

The book details the requirements for filter strips. It advises that they should be between 6 m and 15 m wide. A narrower strip is suggested when used in conjunction with filter or infiltration trenches.

The hydraulic design is based on the use of Manning's roughness to determine velocity and depth of flow. It is advised that velocities should be less than 0.3 m/s to settle out sediments and less than 1.5 m/s to avoid erosion problems. Uneven ground and areas with shrubs will need lower velocities due to the tendency for local low points to be a focus for erosion.

Filter strips have obvious ecological as well as aesthetic benefits.

The choice of vegetation is important because of the salts and high pollution concentrations carried in road washoff.

G15.4 Maintenance requirements

The greatest concerns are the need to maintain the uniform, level surface required and preventing rivulets from forming. Sediments and other debris that gets trapped in filter strips can result in raised areas, which, if left unchecked, can significantly lower the strip's treatment efficiency. Regular monitoring is advised.

G16 Wetlands and constructed reedbeds

G16.1 Description and purpose

A wetland comprises a constructed system of shallow pools that create growing conditions suitable for emergent and riparian wetland plants, explicitly designed to lessen the impacts of stormwater quality and quantity in urban areas. Wetlands are designed and installed to maximise pollutant removal and create wetland habitat through the creation of a matrix of water, sediment, plants and detritus. These collectively provide temporary storage of urban stormwater runoff and remove pollutants from it through a series of complementary physical, chemical and biological pathways. A typical wetland is shown in Figure G10.

G16.2 Design principles

To capture and effectively treat the runoff produced by 90 per cent of storms (which is defined as the design criteria), the wetland should be designed to meet the following basic sizing criteria:

- minimum treatment volume to capture and treat an amount equal to 15 mm of rainfall
- surface area requirement defined as the minimum wetland surface area in proportion to the area being drained is as follows:
  - shallow marsh wetland 2 per cent
  - pond/wetlands 1 per cent
- the dry weather flow path length should be at least twice the width of the wetland
- the inflow and groundwater inputs should be greater than infiltration and evaporation water losses for all designs.

The treatment efficiency of a wetland can be increased by carrying out the following:

- increasing the volume of runoff treatment by:
  - capturing a greater percentage of annual runoff volume
  - providing a longer residence time in the wetland for most storm events
- increasing the surface area to volume ratio:
  - increase the total area of the wetland
  - increase the internal structural complexity of the wetland, by adding complex pond shapes and establishing extensive and dense wetland plant cover

- increasing the effective flow path through the wetland by:
  - extending the distance between the inlet and outlet
  - maximising sinuosity of the dry weather flow path
  - creating areas with an extremely shallow flow path
  - using multiple ponds within the wetland system
- providing runoff pre-treatment and energy dissipation by using a forebay or pond near the inlet, with broad-crested weirs to spread flow. Outlet settlement ponds before the final discharge to the watercourse can also be used
- provide extended detention to keep pollutant removal rates reliable during non-growing season, or utilise a permanent pool to increase algae uptake and sedimentation
- construct a sub-surface flow reedbed as part of the wetland.

Other aspects that should be taken into consideration include establishing the plant community by transplanting stock native to the region. The habitat diversity should be planned to meet the feeding, breeding/nesting and other requirements of a wide range of aquatic, avian and terrestrial species.

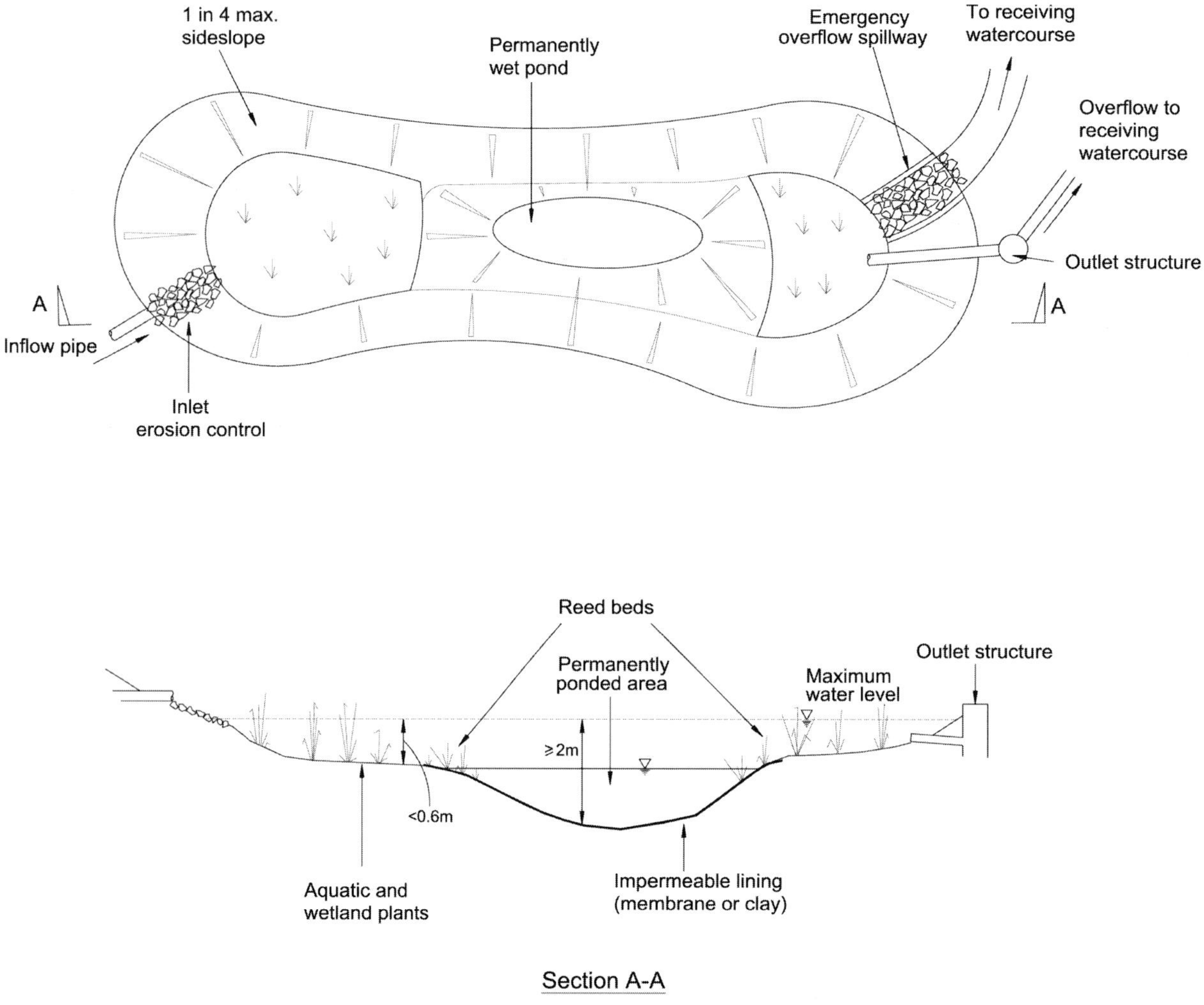

Figure G10 *Wetland*

G16.3 Design method in CIRIA C522

The guide advises that wetland volumes should be sized on the base of $3 \times V_t$. This is thought to provide the necessary retention to effect adequate treatment. A key feature of wetlands is the presumption of baseflow. This differentiates it from many other SUDS units that operate effectively on intermittent inflows.

It is stressed that wetlands are eco-systems and their design must therefore consider all aspects of hydraulic, vegetation, wildlife, climate and chemical processes.

In principle, all the features that need to be considered in building retention ponds are to be considered for wetland ponds. However, to achieve a balanced eco-system, the design of these units is likely to require input from a range of environmental experts.

G16.4 Maintenance requirements

Both the initial establishment and future development of a wetland requires active management of the hydrology and vegetation, as it grows in biomass, diversity and spatial coverage. The design team must plan for the future operation and maintenance of the wetland, with a strong emphasis on the first three years. Maintenance activities must be fully vested with a responsible party through an enforceable maintenance agreement. The agreement should specifically include a projected schedule for inspections and sediment clean-outs, and show evidence that dedicated funding will be available to perform this function. The management agreement would normally be in the form of an environmental management plan to ensure objectives are met.

The wetland should be inspected twice a year in the first three years after construction, with an annual inspection thereafter. Inspections should be conducted with the as-built and landscaping plans, and should take specific note of species distribution/survival, sediment accumulation, water elevations and condition of the outlet. Records should be stored so that the progressive development of the wetland system over time can be monitored.

Accumulated sediment in the wetland should be cleaned out every three to five years. This may involve the draining of the wetland. Silt traps will need cleaning out on more frequently, determined by inspection. The preferred disposal method is on-site land application at a pre-designated spoil area. Some parts of the wetland may require mowing once a year. However, all the remaining areas can be managed as a wetland meadow. Additional information is available in CIRIA Report 180, *Review of the design and management of constructed wetlands* (Nuttall *et al*, 1997).

REFERENCES

Author listing

ABBOTT, C L, CAMINO, L and ANGOOD, C, 2000
Monitoring performance of infiltration drainage systems
Report SR 569, HR Wallingford, Wallingford

ACKER, P, 1958
Resistance of fluids flowing in channels and pipes
Hydraulics Research Paper No 1, HMSO, London

ACKERS, J C, BUTLER, D and MAY, R W P, 1994
Design of sewers to control sediment problems
Report 141, CIRIA, London

AMERICAN PETROLEUM INSTITUTE, 1969
Ch 3 "Oil/water separator process design", in *Manual on disposal of refinery wastes*
American Petroleum Institute, New York

AUCKLAND REGIONAL COUNCIL, 2002
Stormwater management devices: design guidelines manual
Technical Publication 10, Auckland Regional Council, Auckland

BETTESS, R, 1996
Infiltration drainage – manual of good practice
Report 156, CIRIA, London

BRE, 1991
Soakaway design
Digest 365, BRE, Watford

CENTRE FOR ECOLOGY AND HYDROLOGY, 1999
Flood estimation handbook
CEH, Wallingford

CIRIA, 1998
Septic tank systems: a regulator's guide
Special Publication 144BT, CIRIA, London

COLEBROOK, G F, 1939
"Turbulent flow in pipes with particular reference to the transition region between the smooth and rough pipe laws"
J instn civ engrs, vol 11, pp 133–156

COPAS, B A, 1957
"Stormwater storage calculation"
J inst public health engg, vol 56, no 3, pp 137–162

DETR, 1998
Planning for sustainable development: towards better practice
Stationery Office, London

DETR, 1999
Planning requirements in respect of the use of non-mains sewerage incorporating septic tanks in new development
Circular 03/99, Stationery Office, London

DETR, 2000
Housing
PPG3, Stationery Office, London

DoE, 1992
Development and flood risk (superseded by DTLR PPG25)
C30/92, Stationery Office, London

DoE, 1994
Planning and pollution control
PPG23, Stationery Office, London

DoE (Northern Ireland), 1999
The Building Regulations (Northern Ireland) 1994 Amendments booklet
AMD1, Stationery Office, London

DoE, BAKER ASSOCIATES, UNIVERSITY OF THE WEST OF ENGLAND, 1993
Environmental appraisal of development plans: a good practice guide
HMSO, London

DTLR, 2001
Development and flood risk
PPG25 (replacing DoE 30/92), Stationery Office, London

ELLIS, J B and REVITT, M, 1991
Drainage from roads – control and treatment of highway runoff
NRA 43804, UPRC Middlesex University for Thames Region NRA, London

ENVIRONMENT AGENCY, 1996
Environmental assessment: scoping handbook for projects
HMSO, London

ENVIRONMENT AGENCY, 1997a
Liaison with local planning authorities
Environment Agency, Bristol

ENVIRONMENT AGENCY, 1997b
Policy and practice for the protection of flood plains
Environment Agency, Bristol

ENVIRONMENT AGENCY, 1998a
Policy and practice for the protection of groundwater
Stationery Office, London

ENVIRONMENT AGENCY, 1998b
The use and design of oil separators in surface water drainage systems
PPG03, Environment Agency, Bristol

ENVIRONMENT AGENCY, SEPA, ENVIRONMENT AND HERITAGE SERVICE, 2000
Preventing pollution on industrial sites
PPG11, Environment Agency, Bristol

ENVIRONMENT AGENCY, ELLIS, J B and SHUTES, R B E, 2003
Guidance manual for constructed wetlands
R&D Technical Report P2-159/TR1, Environment Agency, Bristol

ENVIRONMENT AGENCY and KELLAGHER, R, 2004
Preliminary rainfall runoff management for developments
R&D Technical Report W5-074/A, Environment Agency, Bristol

ESCARAMEIA, M, GASOWSKI, Y, MAY, R W P and LO CASCIO, A, 2001
Hydraulic design of drainage channels with lateral inflow
Report SR 581, HR Wallingford, Wallingford

ESCARAMEIA, M, GASOWSKI, Y, MAY, R W P and BERGAMINI, L, 2002
Hydraulic design of paved areas
Report SR 606, HR Wallingford, Wallingford

ESCARAMEIA, M and LAUCHLAN, C, 2003
Implications for site drainage design of low water usage in domestic buildings
Report SR 632, HR Wallingford, Wallingford

FORTY, E J, 1998
Performance of gully pots for road drainage
Report SR 508, HR Wallingford, Wallingford

HALL, M J, HOCKIN, D L and ELLIS, J B, 1993
Design of flood storage reservoirs
Book 14, CIRIA, London

HIGHWAYS AGENCY, 1996
Design manual for roads and bridges, vol 4 sec 2, pt 1, Design of outfalls for surface water channels
Advice Note HA 78/96, Stationery Office, London

HIGHWAYS AGENCY, 1997
Design manual for roads and bridges, vol 4 sec 2, pt 4, Hydraulic design of road-edge surface water channels
Advice Note HA 37/97, Stationery Office, London

HIGHWAYS AGENCY, 2000
Design manual for roads and bridges, vol 4 sec 2, pt 3, Spacing of road gullies
Advice Note HR 102/00, Stationery Office, London

HR WALLINGFORD, 1982
Velocity equations for hydraulic design of pipes
HR Wallingford, Wallingford

HR WALLINGFORD and BARR, D I H, 1998
Tables for hydraulic design of pipes, sewers and channels, vol 1, 7th edn
Thomas Telford, London

HR WALLINGFORD and INSTITUTE OF HYDROLOGY, 1981a
The Wallingford Procedure: design and analysis of urban storm drainage
HR Wallingford, Wallingford

HR WALLINGFORD and INSTITUTE OF HYDROLOGY, 1981b
The Wallingford Procedure vol 4, Modified Rational Method
HR Wallingford, Wallingford

INSTITUTION OF CIVIL ENGINEERS, 1975
Guide to the Reservoirs Act 1975
Instn Civ Engrs, London

INSTITUTION OF CIVIL ENGINEERS, 1996
Land drainage and flood defence responsibilities, 3rd edn
Instn Civ Engrs, London

INSTITUTE OF HYDROLOGY, 1975
Flood studies report
Institute of Hydrology, Wallingford

INSTITUTE OF HYDROLOGY, 1978
Flood prediction for small catchments
Flood Studies Supplementary Report 6, Institute of Hydrology, Wallingford

INSTITUTE OF HYDROLOGY, 1983
Review of regional growth curves
Flood Studies Supplementary Report 14, Institute of Hydrology, Wallingford

INSTITUTE OF HYDROLOGY, 1985
The FSR rainfall runoff model parameter estimation equation updated
Flood Studies Supplementary Report 16, Institute of Hydrology, Wallingford

INSTITUTE OF HYDROLOGY, 1994
Flood estimation for small catchments
Report 124, Institute of Hydrology, Wallingford

JOHNSON, R A, LEEK, D S and COPE, M P, 1995
Water-resisting basement construction – summary report. Safeguarding new and existing basements against water and dampness
Report 140, CIRIA, London

KELLAGHER, R, 2000
The Wallingford Procedure for Europe: best practice guide to urban drainage modelling (CD)
WP 6, HR Wallingford, Wallingford

KELLAGHER, R, 2002a
Guide to storage requirements for rainfall runoff from green field development sites
Report SR 580, HR Wallingford, Wallingford

KELLAGHER, R, 2002b
Overview summary – storage requirements for rainfall runoff from green field development sites
Report SR 591, HR Wallingford, Wallingford

LAMONT, P A, 1954
"A review of pipe friction data and formulae, with a proposed set of experimental formulae based on the theory of roughness"
Proc instn civ engrs, vol 3, pt 3, pp 248–277

LEGGETT, D J, BROWN, R, STANFIELD, G, BREWER, D and HOLLIDAY, E, 2001
Rainwater and greywater use in buildings: decision-making for water conservation
Project Report 80, CIRIA, London

LLOYD-DAVIES, D E, 1906
"The elimination of stormwater from sewerage systems"
Proc inst civ engrs, vol 164, pt 2, pp 41–67

MAFF, 1980
Pipe size design for field drainage
Report 5, MAFF, London

MAFF, 1981
The design of field drainage pipe systems
Report 345, MAFF, London

MAGENIS, S, 1994
Design and operation of trash screens
R&D Note 300, National Rivers Authority, London

MARTIN, P, *et al*, 2000a
Sustainable urban drainage systems – design manual for Scotland and Northern Ireland
C521, CIRIA, London

MARTIN, P, *et al*, 2000b
Sustainable urban drainage systems – design manual for England and Wales
C522, CIRIA, London

MARTIN, P, *et al*, 2001
Sustainable urban drainage systems – best practice manual
C523, CIRIA, London

MASKELL, A D (ed), 1992
Scope for control of urban runoff. Vol 1: Overview
Report 123, CIRIA, London

MASKELL, A D, SHERRIFF, J D F and LEONARD, O J (eds), 1992
Scope for control of urban runoff
Vol 2: A review of present methods and practice
Vol 3: Guidelines
Vol 4: A review of legislation, procedures, economic and planning issues
Report 124, CIRIA, London

MAY, R W P, 1982
Design of gutters and gutter outlets: theory and experiment
Report IT 205, HR Wallingford, Wallingford

MAY, R W P, 1984
"Hydraulic design of roof gutters"
Proc instn civ engrs, pt 2, vol 77, pp 479–489

MAY, R W P, 1996
Manual for the hydraulic design of roof drainage systems. A guide to the use of BS 6367:1983
Report SR 485, HR Wallingford, Wallingford

MAY, R W P, MARTIN, P and PRICE, N J, 1998
Low-cost options for prevention of flooding from sewers
C506, CIRIA, London

MINISTRY OF HOUSING AND LOCAL GOVERNMENT, 1970
Technical committee on storm overflows and the disposal of storm sewage
Circular 12/70, HMSO, London

MULVANEY, T J, 1850
"On the use of self registering rain and flood gauges in making observations on the relation of rainfall and flood discharges in a given catchment"
Trans instn civ engrs Ireland, vol 4, no 2, p 12

NATIONAL SUDS WORKING GROUP, 2004
Interim code of practice for SUDS
National SUDS Working Group, Bristol

NEW YORK STATE DEPARTMENT OF ENVIRONMENTAL CONSERVATION, 2001
Stormwater management design manual
Center for Watershed Protection, New York

NUTTALL, P M, BOON, A G and ROWELL, M R, 1997
Review of the design and management of constructed wetlands
Report 180, CIRIA, London

OFWAT, 1992
First time rural sewerage
Information Note 11 (rev Sep 2001), OFWAT, Birmingham

OSBORNE, M, BUTLER, D, CLARK, P and MEMON, F, 1998
Management of gully pots for improved runoff quality
Report 183, CIRIA, London

OVE ARUP & PARTNERS, rev GILBERTSON, A, 2004
CDM Regulations – work sector guidance for designers, 2nd edn
C604, CIRIA, London

PRATT, C, WILSON, S and COOPER, P, 2002
Source control using constructed pervious surfaces: hydraulic, structural and water quality performance issues
C582, CIRIA, London

PRIVETT, K D, MATTHEWS, S C and HODGES, R A, 1996
Barriers, liners and cover systems for containment and control of land contamination
Special Publication 124, CIRIA, London

PROPERTY SERVICES AGENCY, 1987
Technical guide for the prevention of pollution by oil
Civil Engineering Technical Guide no 41, Department of the Environment, Directorate of Civil Engineering Services, London

PRUDHOE, J and YOUNG, C P, 1973
The estimation of flood flows from natural catchments
Report LR565, Transport and Road Research Laboratory, Crowthorne

SCHUELER, T R, 1987
Controlling urban runoff: a practical manual for planning and designing urban BMPs
Metropolitan Washington Council of Governments, Washington DC

SHAFFER, P, ELLIOTT, C, REED, J, HOLMES, J and WARD, M, 2004
Model agreements for sustainable water management systems. Model agreements for SUDS
C625, CIRIA, London

TRRL, 1976
A guide for engineers to the design of storm sewer systems
Road Note 35, Transport and Road Research Laboratory, Crowthorne

TRRL, 1984
The drainage capacity of BS road gullies and a procedure for estimating their spacing
Contractor Report 2, Transport and Road Research Laboratory, Crowthorne

UKWIR, KELLAGHER, R *et al*, 2004
Climate change and the hydraulic design of sewerage systems
Reports 1–15, R&D Project CL/10, UKWIR, London

WILSON, S, BRAY, R and COOPER, P, 2004
Sustainable drainage systems. Hydraulic, structural and water quality advice
C609, CIRIA, London

WRC, 2001
Sewers for adoption. A design and construction guide for developers, 5th edn
WRc, Swindon

YOUNG, O C and O'REILLY, M P, 1987
A guide to design loadings for buried rigid pipes, 2nd edn
HMSO, London

British Standards

BS

BS 6297:1983 *Code of practice for design and installation of small sewage treatment works and cesspools*

BS 6367:1983 *Drainage of roofs and paved areas*
Superseded by BS EN 12056-3:2000

BS 8005-1:1987 *Sewerage. Guide to new sewerage construction*
Superseded by BS EN 752-4:1998

BS 8301:1985 *Code of practice for building drainage*

BS EN

BS EN 124:1994 *Gully tops and manhole tops for vehicular and pedestrian areas*

BS EN 476:1997 *General requirements for components used in discharge pipes, drains and sewers for gravity systems*

BS EN 752-2:1997 *Drain and sewer systems outside buildings. Performance requirements*

BS EN 752-3:1997 *Drain and sewer systems outside buildings. Planning*

BS EN 752-4:1998 *Drain and sewer systems outside buildings. Hydraulic design and environmental considerations*

BS EN 858-1:2002 *Separator systems for light liquids (eg oil and petrol). Principles of product design, performance and testing, making and quality control*

BS EN 1433:2002 *Drainage channels for vehicular and pedestrian areas. Classification, design and testing requirements, marking and evaluation of conformity*

BS EN 12050:2001 *Wastewater lifting plants for buildings and sites – principles of construction and testing*

BS EN 12056-3:2000 *Gravity drainage systems inside buildings. Roof drainage, layout and calculation*

BS EN 12056-4:2000 *Gravity drainage systems inside buildings. Wastewater lifting plants for buildings and sites – layout and calculation*

BS EN 12109:1999 *Vacuum drainage systems inside buildings*

EN

EN 1825-2:2002 *Grease separators. Selection of nominal size, installation, operation and maintenance*

EN 12056-3:2000 *Gravity drainage systems inside buildings. Roof drainage, layout and calculation*

Legislation etc

BUILDING ACT 1984 [or 1989, as original text, but not on TSO website]
1984 c. 55, HMSO, London

THE BUILDING REGULATIONS (ENGLAND & WALES) 2000 (and subsequent amendments)
Schedule 1, Part H Drainage and waste disposal
SI 2000/2531, 2001/3335, 2002/402, 2002/2871, 2003/2692, Stationery Office, London

BUILDING REGULATIONS (NORTHERN IRELAND) 2000
Statutory Rule of NI 2000/389, Stationery Office, London

BUILDING STANDARDS (SCOTLAND) AMENDMENT REGULATIONS 2001
Scottish Statutory Instrument 2001/320, Stationery Office, London

ENVIRONMENT ACT 1995
1995 c. 25, HMSO, London

THE GROUNDWATER REGULATIONS 1998
SI 1998/2746, Stationery Office, London

HIGHWAYS ACT 1980
1980 c. 66, HMSO, London

LAND DRAINAGE ACT 1991
1991 c. 59, HMSO, London

LOCAL GOVERNMENT ACT 1972
1972 c. 70, HMSO, London

LOCAL GOVERNMENT ETC. (SCOTLAND) ACT 1994
1994 c. 39, HMSO, London

PUBLIC HEALTH ACT 1936
26 Geo V & 1 Edw VIII c. 49, HMSO, London

RESERVOIRS ACT 1975
1975 c. 23, HMSO, London

TOWN AND COUNTRY PLANNING ACT 1990
1990 c. 8, HMSO, London

TOWN AND COUNTRY PLANNING ORDER 1995
SI 1995/419, HMSO, London

TRANSPORT ACT 1962
1962 c. 46, HMSO, London

TRANSPORT ACT 1968
1968 c. 73, HMSO, London

WATER ACT 1989
1989 c. 15, HMSO, London

WATER INDUSTRY ACT 1991
1991 c. 56, HMSO, London

WATER RESOURCES ACT 1991
1991 c. 57, HMSO, London

Report listing (by publisher)

Auckland Regional Council

Technical Publication 10 *Stormwater management devices: design guidelines manual* (Auckland Regional Council, 2002)

BRE

Digest 365 *Soakaway design* (BRE, 1991)

CIRIA

B14 *Design of flood storage reservoirs* (Hall *et al*, 1993)

C506 *Low-cost options for the prevention of flooding from sewers* (May, 1998)

C521 *Sustainable urban drainage systems – design manual for Scotland and Northern Ireland* (Martin *et al*, 2000a)

C522 *Sustainable urban drainage systems – design manual for England and Wales* (Martin *et al*, 2000b)

C523 *Sustainable urban drainage systems – best practice manual* (Martin *et al*, 2001)

C582 *Source control using constructed pervious surfaces: hydraulic, structural and water quality performance issues* (Pratt *et al*, 2002)

C604 *CDM Regulations – work sector guidance for designers*, 2nd edn (Ove Arup and Gilbertson, 2004)

C609 *Sustainable drainage systems. Hydraulic, structural and water quality advice* (Wilson *et al*, 2004)

PR80 *Rainwater and greywater use in buildings: decision-making for water conservation* (Leggett *et al*, 2001)

R123 *Scope for control of urban runoff. Vol 1: Overview* (Maskell, 1992)

R124 *Scope for control of urban runoff. Vol 2: A review of present methods and procedures; Vol 3: Guidelines; Vol 4: A review of legislation, procedures, economic and planning issues* (Maskell *et al*, 1992)

R140 *Water-resisting basement construction – summary report. Safeguarding new and existing basements against water and dampness* (Johnson *et al*, 1995)

R141 *Design of sewers to control sediment problems* (Ackers *et al*, 1994)

R156 *Infiltration drainage – manual of good practice* (Bettess, 1996)

R180 *Review of the design and management of constructed wetlands* (Nuttall *et al*, 1997)

R183 *Management of gully pots for improved runoff quality* (Osborne *et al*, 1998)

SP124 *Barriers, liners and cover systems for containment and control of land contamination* (Privett *et al*, 1996)

SP144BT *Septic tank systems: a regulator's guide* (CIRIA, 1998)

DoE, DETR, DTLR

C30/92 *Development and flood risk* (superseded by PPG25) (DoE, 1992)

03/99 *Planning requirements in respect of the use of non-mains sewerage incorporating septic tanks in new development* (DETR, 1999)

PPG3 *Housing* (DETR, 2000)

PPG23 *Planning and pollution control* (DoE, 1994)

PPG25 *Development and flood risk* DTLR, (2001)

Environment Agency

PPG03 *The use and design of oil separators in surface water drainage systems* (EA, 1998b)

PPG11 *Preventing pollution on industrial sites* (EA, 2000)

P2-159/TR1 *Guidance manual for constructed wetlands* (EA, 2003)

Highways Agency

HA 78/96 *DMRB, v 4, s 2, pt 1 Design of outfalls for surface water channels* (HA, 1996)

HA 37/97 *DMRB, v 4, s 2, pt 4 Hydraulic design of road-edge surface water channels* (HA, 1997)

HR 102/00 *DMRB, v 4, s 2, pt 3 Spacing of road gullies* (HA, 2000)

HR Wallingford

IT 205 *Design of gutters and gutter outlets: theory and experiment* (May, 1982)

SR 485 *Manual for the hydraulic design of roof drainage systems. A guide to the use of BS 6367:1983* (May, 1996)

SR 508 *Performance of gully pots for road drainage* (Forty, 1998)

SR 569 *Monitoring performance of infiltration drainage systems* (Abbott *et al*, 2000)

SR 580 *Guide to storage requirements for rainfall runoff from green field development sites* (Kellagher, 2002a)

SR 581 *Hydraulic design of drainage channels with lateral inflow* (Escarameia *et al*, 2001)

SR 591 *Overview summary – storage requirements for rainfall runoff from green field development sites* (Kellagher, 2002b)

SR 606 *Hydraulic design of paved areas* (Escarameia *et al*, 2002)

SR 632 *Implications for site drainage design of low water usage in domestic buildings* (Escarameia and Lauchlan, 2003)

WP 6 *The Wallingford Procedure for Europe: best practice guide to urban drainage modelling* (Kellagher, 2000)

Institute of Hydrology

FSSR 6 *Flood prediction for small catchments* (IH, 1978)

FSSR 14 *Review of regional growth curves* (IH, 1983)

FSSR 16 *The FSR rainfall runoff model parameter estimation equation updated* (IH, 1985)

Report 124 *Flood estimation for small catchments* (IH, 1994)

Ministry of Agriculture, Fisheries and Food (MAFF)

Report 5 *Pipe size design for field drainage* (MAFF, 1980)

Report 345 *The design of field drainage pipe systems* (MAFF, 1981)

National Rivers Authority (NRA)

NRA 43804 *Drainage from roads – control and treatment of highway runoff* (Ellis and Revitt, 1991)

R&D Note 300 *Design and operation of trash screens* (Magenis, 1994)

Transport and Road Research Laboratory (TRRL)

Contractor Report 2 *The drainage capacity of BS road gullies and a procedure for estimating their spacing* (TRRL, 1984)

LR565 *The estimation of flood flows from natural catchments* (Prudhoe and Young, 1973)

Road Note 35 *A guide for engineers to the design of storm sewer systems* (TRRL, 1976)